*Numerical Methods
for Solving*
Nonlinear Equations

D. James Benton

Copyright © 2018 by D. James Benton, all rights reserved.

Foreword

Nonlinear equations are found throughout science and engineering across a wide variety of disciplines. These are a significant part of applied mathematics and much effort has been devoted to their study and solution. We will consider single and multi-variable problems as well as real and complex numbers. Various theories will be presented, but always with a focus on what works best—that is, the most practical approaches. In the literature of applied mathematics, these are known as *robust* algorithms.

All of the examples contained in this book,
(as well as a lot of free programs) are available at...
http://www.dudleybenton.altervista.org/software/index.html

Programming

Most of the examples in this book are implemented in the C programming language. A few are implemented in VBA® (Visual BASIC for Applications, or what Microsoft® calls the language of Excel® macros). BASIC stands for Beginner's All-Purpose Symbolic Instruction Code. If you're still using some form of BASIC and haven't yet graduated to a professional programming language, now is the time to do so and there is nothing better than C. It's in a *class* by itself.

Table of Contents

	page
Foreword	i
Programming	i
Chapter 1. Single-Valued Functions of One Variable	1
Chapter 2. Multi-Valued Functions of One Variable	14
Chapter 3. Functions of Several Variables	20
Chapter 4: Bounding & Scaling	28
Chapter 5: Nonlinear Least-Squares	30
Chapter 6: Quasi-Newton Methods	34
Chapter 7. Broyden's Derivative-Free Algorithm	42
Chapter 8. Approximating Gradients and Hessians	46
Chapter 9. Multi-Dimensional Bisection Search	50
Chapter 10: Evolutionary Method	56
Chapter 11: Nonlinear Regression	62
Chapter 12: Hybrid Regression	79
Appendix A: Steam Properties - A Practical Example	81
Appendix B: Graphical Representation	90
Appendix C: SIAM Paper	92

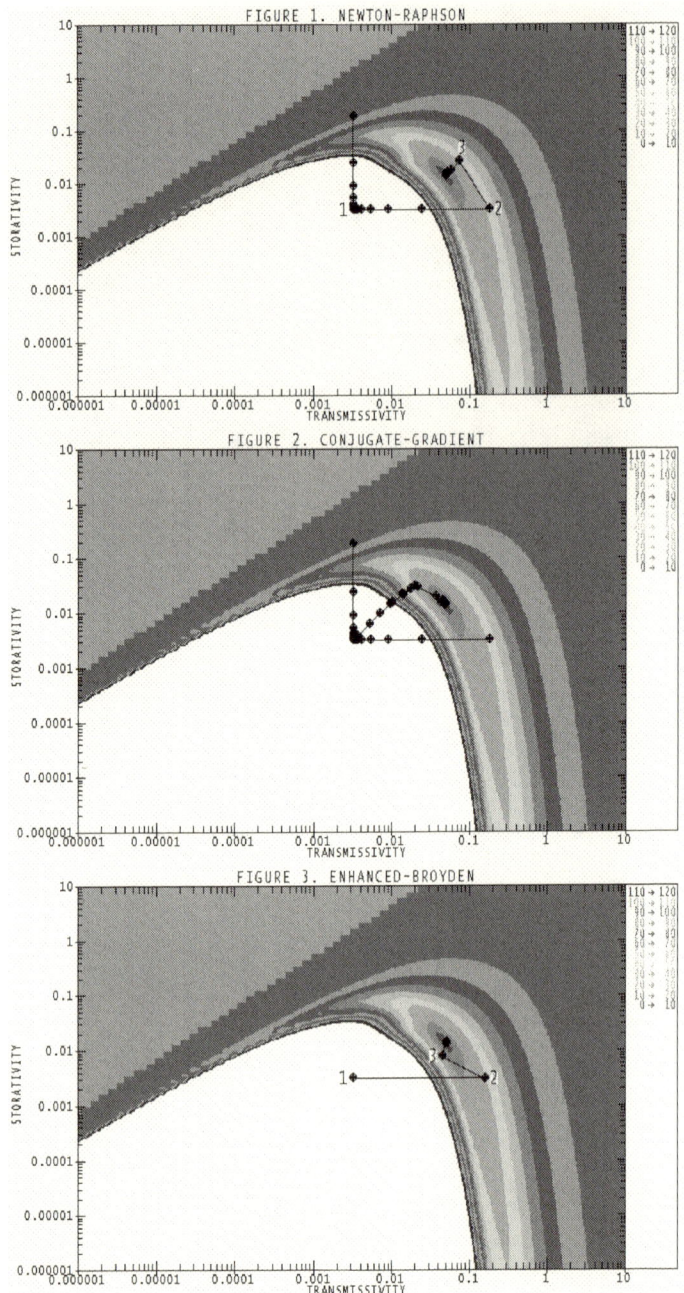

Chapter 1. Single-Valued Functions of One Variable

The simplest problem we can consider is that of single-valued, real functions of one variable—that is, $y(x)$ where y has one-and-only-one value for each x and there is only one value of x for any value of y. This excludes functions that wrap around, fold back on themselves, oscillate, or take on complex values, such as $\sin(x)$, $\cos(x)$, and $\sqrt{1-x^2}$. One such function is illustrated in the following figure:

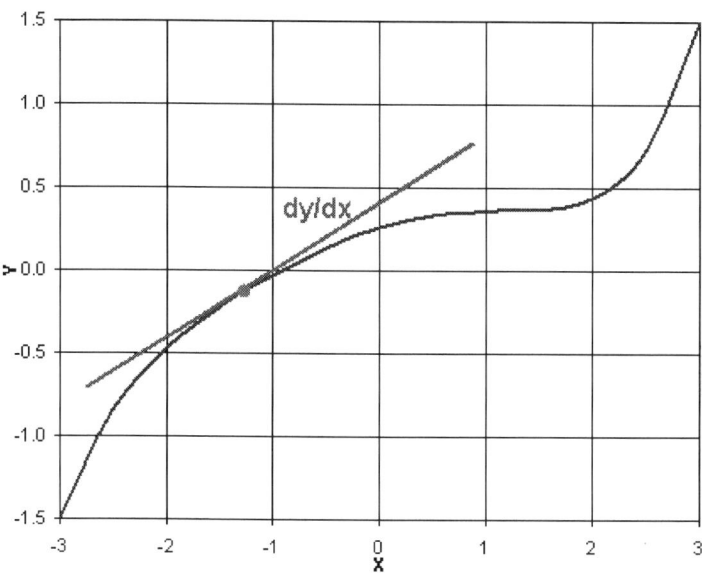

Most discussions presume you're interested in $y(x)=0$. For some other target value, simply subtract $y'(x)=y(x)-target$ and solve the modified problem. We will consider several methods for solving this problem.

Newton-Raphson Method

The Newton-Raphson Method begins with a guess (e.g., the purple dot in the preceding figure), the value of the function at that position, and the slope (or first derivative) of the function at that point.[1] The next guess (or *provisional value*) is calculated from the current one by taking the difference between the target and current value of the function and dividing this by the slope.

---

[1] Named after Isaac Newton and Joseph Raphson, was first published by John Wallis in "A Treatise of Algebra both Historical and Practical," in 1685.

$$x_{n+1} = x_n + \frac{y(x_n)}{\left(\frac{dy}{dx}\right)_{x_n}} \tag{1.1}$$

This method works well some of the time, but not always. It may converge rapidly… or not at all. If the root (i.e., where $y=0$) lies in a region of x where dy/dx is small, changes sign, or vanishes (e.g., near $x=1.33$ in the preceding figure), you've got a real problem, because the iteration implied by Equation 1.1 will diverge (i.e., *blow up*). This is why you never want to use the raw, unqualified Newton-Raphson Method to solve any routine problem—you can't depend on it always arriving at a valid answer. The illustration we will use for this is solving the properties of steam (or a refrigerant) for unknown density, which you may need to do many, many times in a spreadsheet or other model. Here's what the iterations look like in Excel® (sorry about the ratty graphics, but it's the best you can get out of Microsoft®).

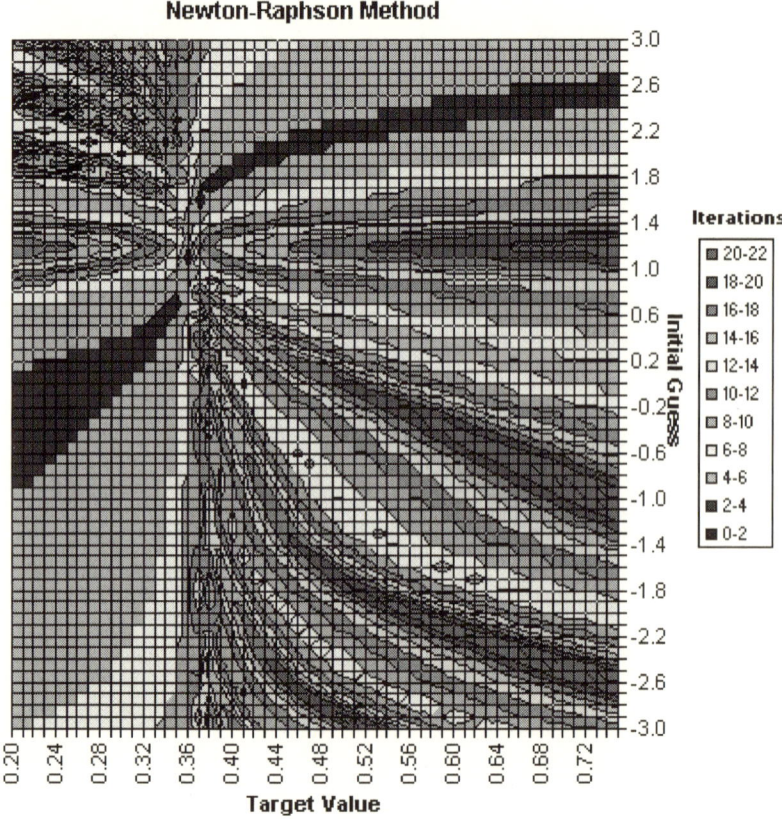

The Newton-Raphson Method always finds a solution with 61x56=3416 cases. The average number of iterations for this example is 9.4. The dark blue curving region (corresponding to zero iterations) is the shape of the curve, y(x), when the initial guess is equal (or very close to) the final solution. The red and magenta regions are where it takes 20-22 iterations to arrive at a solution to the specified tolerance (0.000001). If you're lucky and your first guess is close to the final solution (i.e., in the blue areas) this method is quite efficient. If you're not so lucky (i.e., in the yellow, orange, red, or magenta areas), this method isn't efficient. At least it always converges for this example—something that is not always true. The Newton-Raphson Method in VBA® is:

```
Function NewtonRaphson(target As Double, guess As
   Double, convergence As Double) As Integer
   Dim x As Double
   x = guess
   NewtonRaphson = 1 'counting function calls
   While True
      If (Abs(Y(x) - target) <= convergence) Then Exit
   Function
      NewtonRaphson = NewtonRaphson + 1
      x = x + (target - Y(x)) / dYdX(x)
   Wend
End Function
```

Secant Method

We will next consider the Secant Method. The Newton-Raphson Method required knowledge of the derivative of the function. This is not always available. It may even be quite problematic to estimate the derivative. For example, where would you get the derivative of a stock price on the exchange? In such cases, you only have previous values of the function to work with and must come up with a *derivative-free* algorithm. One of the most basic principles of calculus is:

$$\frac{dy}{dx} \approx \frac{y_2 - y_1}{x_2 - x_1} \qquad (1.2)$$

and so we use this approximation to formulate an iterative step:

$$x_3 = \frac{x_1 y_2 - x_2 y_1}{y_2 - y_1} \qquad (1.3)$$

here the subscripts 1, 2, and 3 indicate the previous two values and the next step, respectively. It should be clear from Equation 1.3 that this method will run into trouble whenever $y_2 \approx y_1$, which happens more often than you might think. You must also come up with something to start this process off (i.e., the first step). The results for this problem are shown in this next figure:

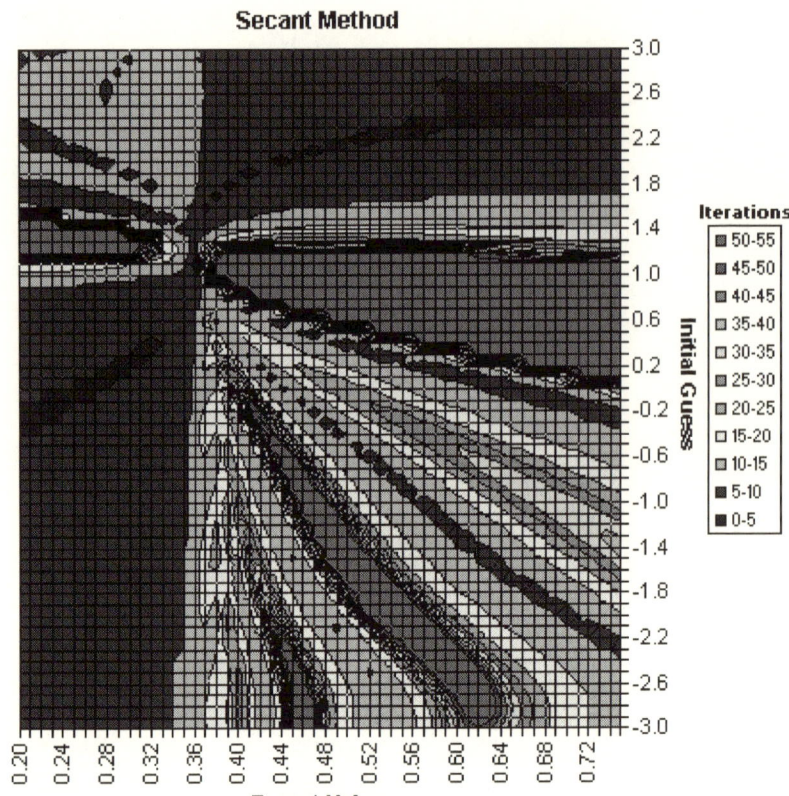

Don't presume that the Secant Method has a larger blue area (sweet spot for convergence) than the Newton-Raphson. In fact, the Secant Method performs so much worse for this problem that the scale has been increased from 22 to 55 iterations. Even worse that that is the fact that the values greater than 54 (i.e., the magenta regions) no solution is found, that is, the method diverges, rather than converges for 13% of the cases. When it did converge the average number of iterations required was 12.9, making the Secant Method a real loser. Its only value is historical. Don't use it for anything. The Secant Method in VBA® is:

```
Function Secant(target As Double, guess As Double,
    convergence As Double) As Integer
  Dim x1 As Double, x2 As Double, x3 As Double, f1 As
    Double, f2 As Double
  x1 = guess
  f2 = Y(x1) - target
  x2 = x1 - f2
  Secant = 2 'counting function calls
  While True
```

```
      f1 = f2
      f2 = Y(x2) - target
      If (Abs(f2) <= convergence Or Secant > 53) Then Exit
   Function
      Secant = Secant + 1
      x3 = (x1 * f2 - x2 * f1) / (f2 - f1)
      x1 = x2
      x2 = x3
   Wend
End Function
```

Regula Falsi Method

The next method we will consider is the Regula Falsi (Latin for *False Position*). Ideally, you want to always bracket the solution, that is, have one point on either side. With Regula Falsi, you may replace either the current or the previous step with the update, to achieve this. Regula Falsi is illustrated in the following figure:

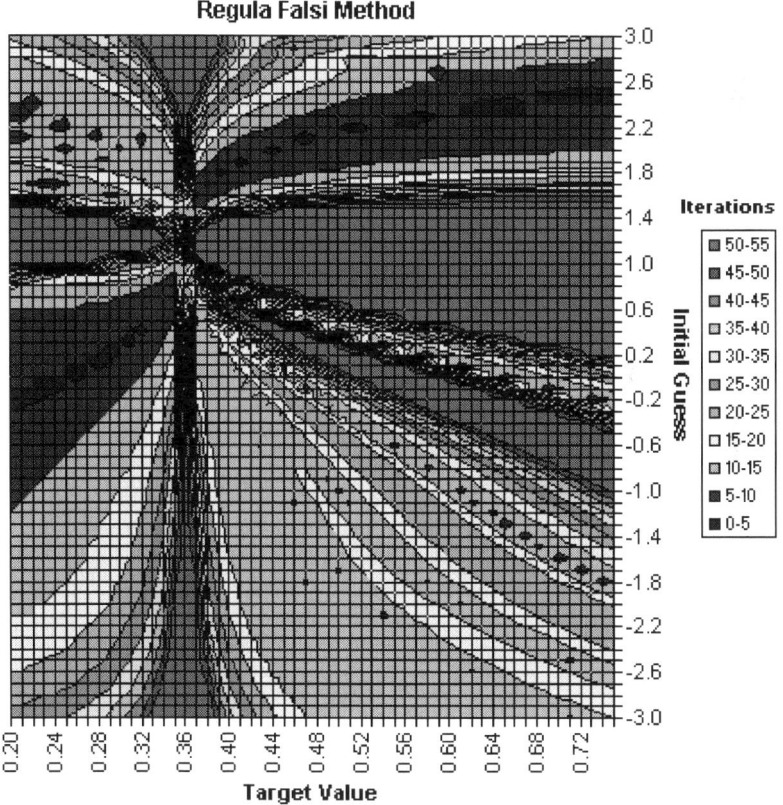

Notice that the regions of non-convergence (magenta areas) in the previous two graphs are different and substantial, making both of these methods a poor choice for applications. This method failed to converge for 19% of the cases and the average number of iterations required to converge when successful was 17.4, making this worse than the Secant Method. Regula Falsi in VBA® is:

```
Function RegulaFalsi(target As Double, guess As Double,
      convergence As Double) As Integer
   Dim x1 As Double, x2 As Double, x3 As Double, f1 As
      Double, f2 As Double, f3 As Double
   x1 = guess
   f1 = Y(x1) - target
   x2 = x1 - f1
   f2 = Y(x2) - target
   RegulaFalsi = 2 'counting function calls
   While True
      If (Abs(f2) <= convergence Or RegulaFalsi > 53) Then
      Exit Function
      RegulaFalsi = RegulaFalsi + 1
      x3 = (x1 * f2 - x2 * f1) / (f2 - f1)
      f3 = Y(x3) - target
      If (f2 * f3 < 0) Then
         x1 = x2
         x2 = x3
         f1 = f2
         f2 = f3
      Else
         x2 = x3
         f2 = f3
      End If
   Wend
End Function
```

Illinois Method

Both the Secant and Regula Falsi methods often overshoot. The Regula Falsi Method may also get bogged down on one side or the other, always updating steps on the same side of the root and significantly slowing—even preventing—convergence. These observations provide motivation for somehow adjusting the updates. One such modification is known as the Illinois Method, published by Ford.[2] There are two variants. The firs½t is:

$$x_3 = \frac{½x_1 y_2 - x_2 y_1}{½y_2 - y_1} \quad (1.4)$$

---

[2] Ford, J. A., "Improved Algorithms of Illinois-Type for the Numerical Solution of Nonlinear Equations," Technical Report CSM-257, University of Essex Press, 1995.

Results for the first variant of this method and the same problem.

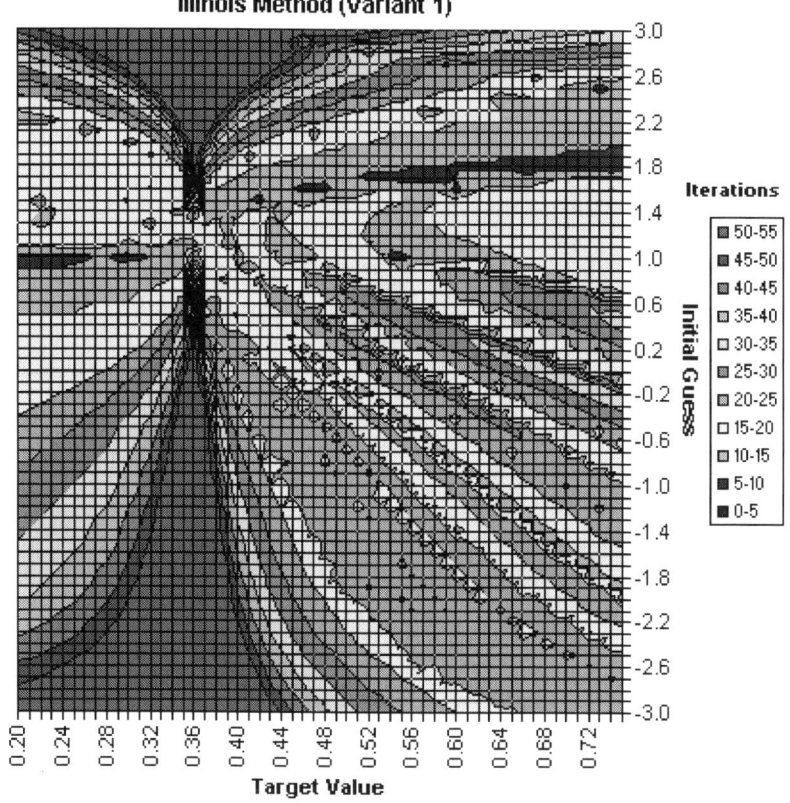

This method failed to converge for 11% of the cases and the average number of iterations required for convergence when successful was 22.8. The first variant in VBA® is:

```
Function Illinois1(target As Double, guess As Double,
    convergence As Double) As Integer
  Dim x1 As Double, x2 As Double, x3 As Double, f1 As
    Double, f2 As Double, f3 As Double
  x1 = guess
  f1 = Y(x1) - target
  x2 = x1 - f1
  f2 = Y(x2) - target
  Illinois1 = 2 'counting function calls
  While True
    If (Abs(f2) <= convergence Or Illinois1 > 53) Then
    Exit Function
    Illinois1 = Illinois1 + 1
```

```
x3 = (x1 * f2 / 2 - x2 * f1) / (f2 / 2 - f1)
f3 = Y(x3) - target
If (f2 * f3 < 0) Then
   x1 = x2
   x2 = x3
   f1 = f2
   f2 = f3
Else
   x2 = x3
   f2 = f3
End If
Wend     End Function
```

The second variant is slightly different:

$$x_3 = \frac{x_1 y_2 - \tfrac{1}{2} x_2 y_1}{y_2 - \tfrac{1}{2} y_1} \tag{1.5}$$

Results for the second variant is shown in this next figure:

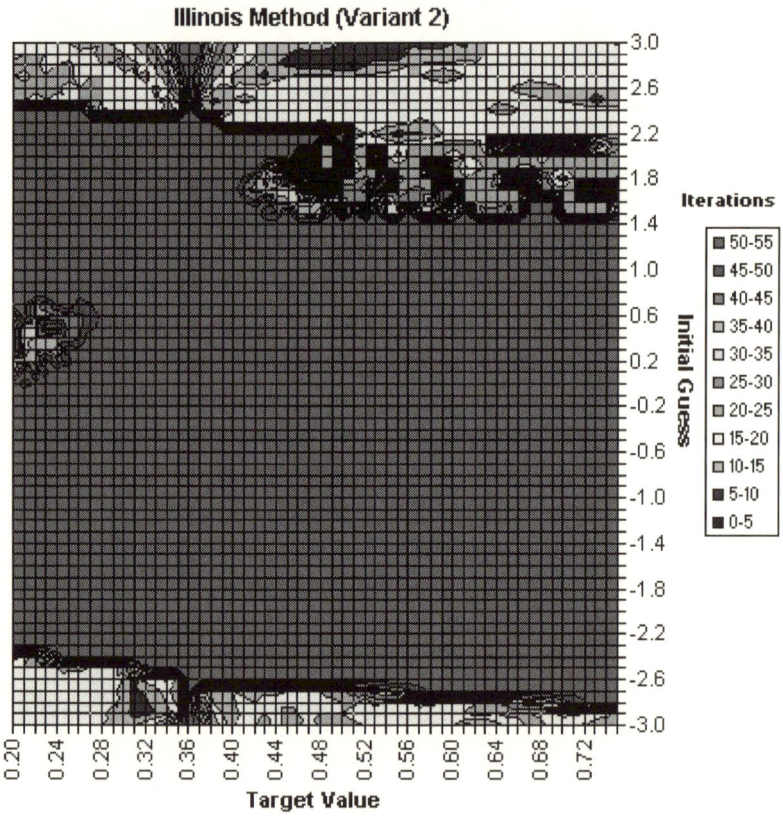

Clearly, the second variant doesn't work at all for this problem so there's no point listing the code that only differs on one line.

Anderson-Björck Method

Another modification has been proposed by Anderson and Björck.[3] This is similar to the Illinois variants, except the ½ is replaced by a factor m, as follows:

$$m = 1 - \frac{y_3}{y_2} \tag{1.6}$$

If m is not positive, then it's set to ½ as before. The results for this method are illustrated in the following figure:

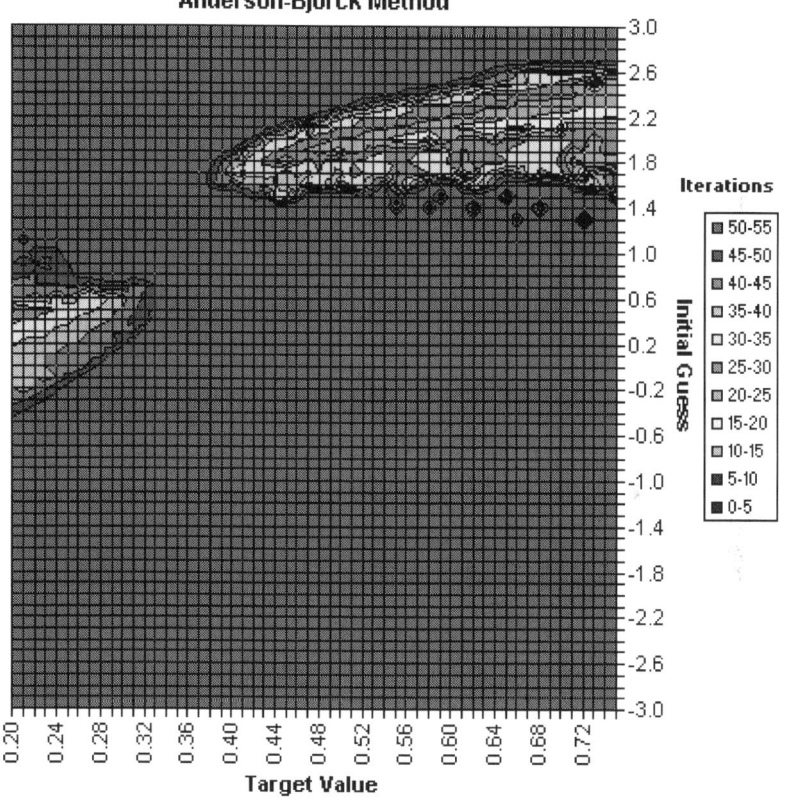

While this method is said to have won a contest, yet it clearly doesn't work for the current problem, failing to converge for 86% of the cases.

---

[3] Dahlquist, G. and Björck, Å., *Numerical Methods*, pp. 231–232, Dover, 1974.

Dekker's Method

Dekker and Hoffmann proposed that the Secant Method be modified to avoid dividing by zero in the case of $y_2 \approx y_1$, by replacing Equation 1.3 with $x_3=(x_1+x_2)/2$.[4] This is somewhat of an improvement, as illustrated in the following figure:

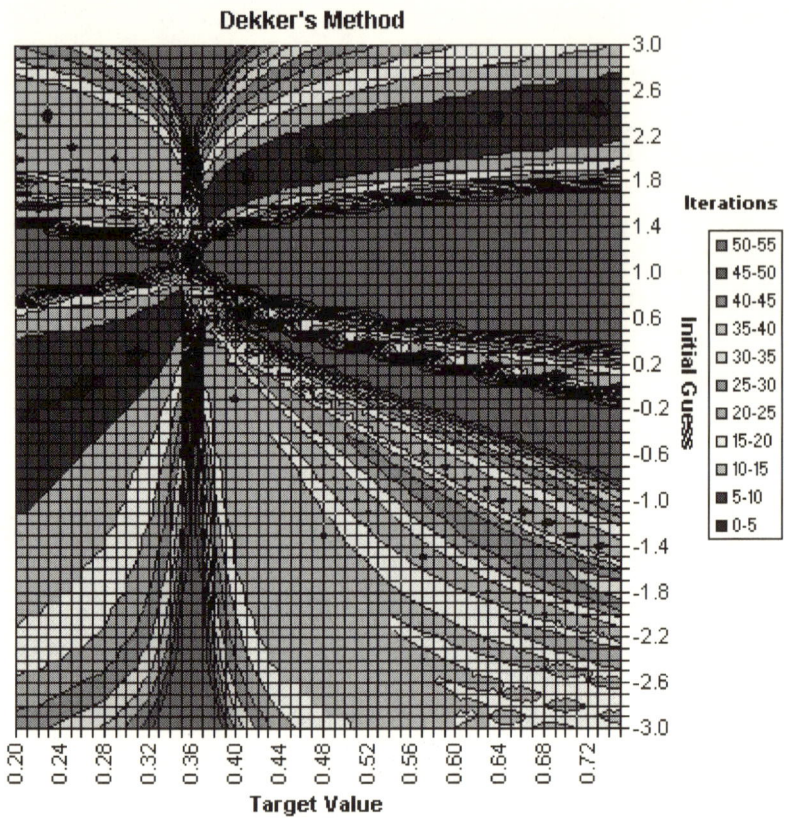

Bisection Search

While there may be some problems where these methods are adequate, for general application, they're useless because not finding a solution is simply not an option! The Newton-Raphson Method can be constrained and tweaked enough to work for some very special problems, and we will present one of those next, but first, the method that makes all of the previous superfluous: bisection search. It works all the time no matter what! It also converges every

---

[4] Dekker, T. J. and Hoffmann, W., "Algol 60 Procedures in Numerical Algebra," Tracts 22 and 23, *Mathematisch Centrum Amsterdam*, 1968.

single time, regardless to $1/2^n$, where n is the number of bisections. It couldn't be any more simple. Of course, it always takes 32 iterations, but that isn't a problem most of the time.

```
Function Bisection(target As Double) As Double
    Dim iter As Integer, x1 As Double, x2 As Double
    x1 = -3
    x2 = 3
    For iter = 1 To 32
        Bisection = (x1 + x2) / 2
        If (Y(Bisection) < target) Then
            x1 = Bisection
        Else
            x2 = Bisection
        End If
    Next iter
End Function
```

A practical application of the Newton-Raphson Method and comparison to the bisection search may be found in Appendix A. This book is not about making pretty pictures. It's about solving practical problems. I have provided a program to make pretty pictures in Appendix B.

Brent's Method

The desired solution is not always a root or zero point. We often seek a minimum or maximum. The preceding methods only work for functions that cross over the y-axis (i.e., have both positive and negative values). Brent's method searches for a minimum.[5] Where the Secant Method arises from a linear interpolant, Brent's method arises from a quadratic interpolant (i.e., the second order Lagrange interpolating polynomial). The iteration begins with three points (x_1, x_2, x_3), presumably bracketing the minimum. The function is evaluated at these three points (y_1, y_2, y_3). The resulting polynomial is:

$$y = \frac{(x-x_2)(x-x_3)y_1}{(x_1-x_2)(x_1-x_3)} + \frac{(x-x_1)(x-x_3)y_2}{(x_2-x_1)(x_2-x_3)} + \frac{(x-x_1)(x-x_2)y_3}{(x_3-x_1)(x_3-x_2)} \quad (1.7)$$

Take the derivative of Equation 1.7 with respect to x, set this to zero, and solve to obtain:

$$x = \frac{1}{2} \frac{(x_2^2-x_3^2)y_1 + (x_3^2-x_1^2)y_2 + (x_1^2-x_2^2)y_3}{y_1(x_2-x_3) + y_2(x_3-x_1) + y_3(x_1-x_2)} \quad (1.8)$$

---

[5] Brent, R. P., *Algorithms for Minimization without Derivatives*, Chapter 4: An Algorithm with Guaranteed Convergence for Finding a Zero of a Function, Prentice-Hall, Englewood Cliffs, NJ, 1973.

If the revised estimate is to the left x_2, swap out x_3; otherwise, swap out x_1. Continue until the improvement or the step size is less than some tolerance. A typical problem is illustrated in the following figure:

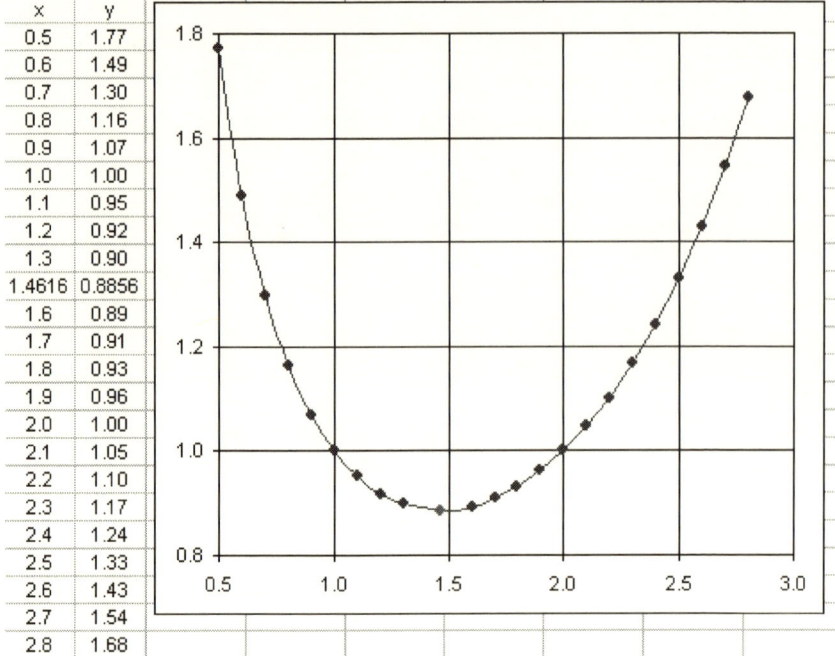

x	y
0.5	1.77
0.6	1.49
0.7	1.30
0.8	1.16
0.9	1.07
1.0	1.00
1.1	0.95
1.2	0.92
1.3	0.90
1.4616	0.8856
1.6	0.89
1.7	0.91
1.8	0.93
1.9	0.96
2.0	1.00
2.1	1.05
2.2	1.10
2.3	1.17
2.4	1.24
2.5	1.33
2.6	1.43
2.7	1.54
2.8	1.68

The algorithm is simple to implement:

```
double Brent(double x1,double x3,double func(double),int
    itmax,double eps,double*x2)
     {
     int iter;
     double x,y1,y2,y3;
     *x2=(x1+x3)/2.;
     y1=func(x1);
     y2=func(*x2);
     y3=func(x3);
     if(y2>=y1||y2>=y3)
        return(y2);
     for(iter=0;iter<itmax;iter++)
         {
         x=y1*(*x2-x3)+y2*(x3-x1)+(x1-*x2)*y3;
         if(fabs(x)<eps*(x3-*x2))
             break;
         x=((*x2*(*x2)-x3*x3)*y1+(x3*x3-x1*x1)*y2+(x1*x1-
     *x2*(*x2))*y3)/x/2.;
         if(x<*x2-eps)
             {
```

```
    x3=*x2;
    y3=y2;
    }
  else if(x>*x2+eps)
    {
    x1=*x2;
    y1=y2;
    }
  else
    break;
  *x2=x;
  y2=func(*x2);
  }
return(y2);
}
```

This example converges in 5 iterations requiring only 8 function calls. The program and spreadsheet can be found in the examples\Brent folder. We will use this same algorithm for step length optimization in subsequent multi-variable problems.

Chapter 2. Multi-Valued Functions of One Variable

We will next consider problems, which have more than one solution; specifically, functions of one variable that have more than one zero. This is basically the same as the first problem, only broken up into intervals.

Bessel Function Zeroes

We begin by finding the zeroes of Bessel functions. Why? Because Bessel functions come with Excel® and finding the zeroes of cos(x) or sin(x) seems pretty silly. The first 4 Bessel functions of the second kind (J_0, J_1, J_2, and J_3) are:

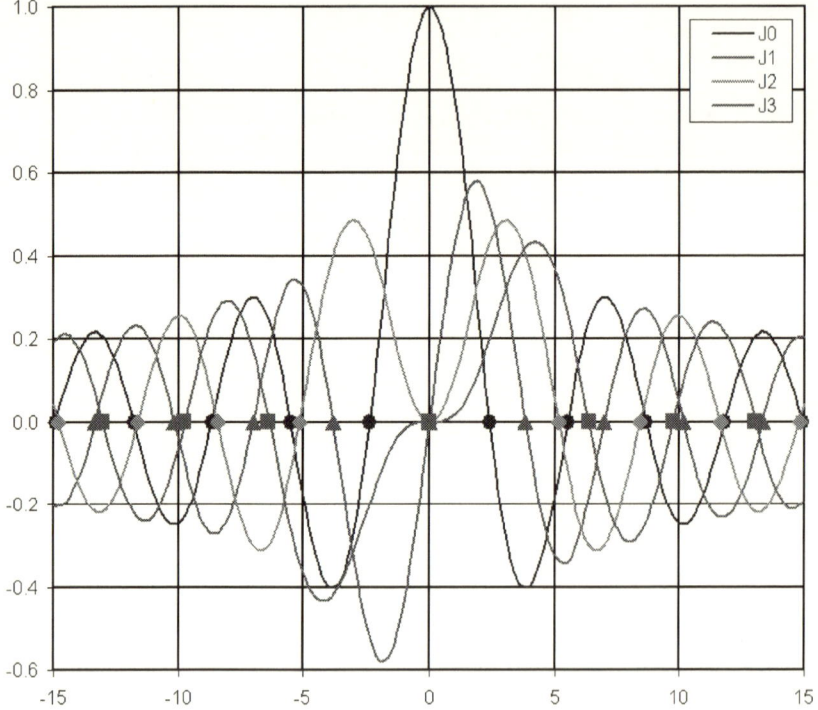

The roots are also shown in the figure by spots of the same color as the curves. These are easily found using a bisection search so that the derivative is not required. The simplest way to accomplish this is through a re-entrant subroutine (i.e., a function that calls itself). We start with the interval -15 to +15 and keep splitting it in half until each interval contains a single root, then we find that root using a bisection search.

```
Sub FindRoot(n As Integer, ByVal x1 As Double, ByVal x2
    As Double, l As Integer)
  If (l > 0) Then
    Call FindRoot(n, x1, (x1 + x2) / 2, l - 1)
```

```
      Call FindRoot(n, (x1 + x2) / 2, x2, 1 - 1)
      GoTo the_end
  End If
  Dim i As Integer, b1 As Double, b2 As Double, x As
    Double
  b1 = Application.Run("BESSELJ", x1, n)
  b2 = Application.Run("BESSELJ", x2, n)
  If (b1 * b2 > 0#) Then GoTo the_end
  For i = 1 To 32
    x = (x1 + x2) / 2
    If (Application.Run("BESSELJ", x, n) < 0) Then
      If (b1 < b2) Then
        x1 = x
      Else
        x2 = x
      End If
    ElseIf (b1 < b2) Then
      x2 = x
    Else
      x1 = x
    End If
  Next i
  For i = 2 To 11
    If (IsEmpty(Cells(i, n + 7))) Then
      Cells(i, n + 7).Value = x
      Exit For
    End If
  Next i
the_end:
End Sub
Private Sub CommandButton1_Click()
  Range("G2:J12").Select
  Selection.ClearContents
  Dim n As Integer
  For n = 0 To 3
    Call FindRoot(n, -15, 15, 4)
  Next n
End Sub
```

Notice that you must specify *byVal* (i.e., by *value* in contrast to by *reference*) in the function declaration so that the value of x1 and x2 are only changed locally. Note also that re-entrant code must always have a mechanism to end the cycle; otherwise, it will get deeper and deeper until you run out of memory—typically resulting in a stack overflow. The files can be found in the online archive in the folder examples\Bessel.

Legendre Polynomial Roots

We will next consider a very useful root finding procedure and one that employs the Newton-Raphson Method: Legendre polynomials. The reason this is useful is that the roots are the abscissas for Gauss Quadrature, the premier method of numerical integration. Graphically, this problem is quite similar to the previous one, as illustrated in this next figure:

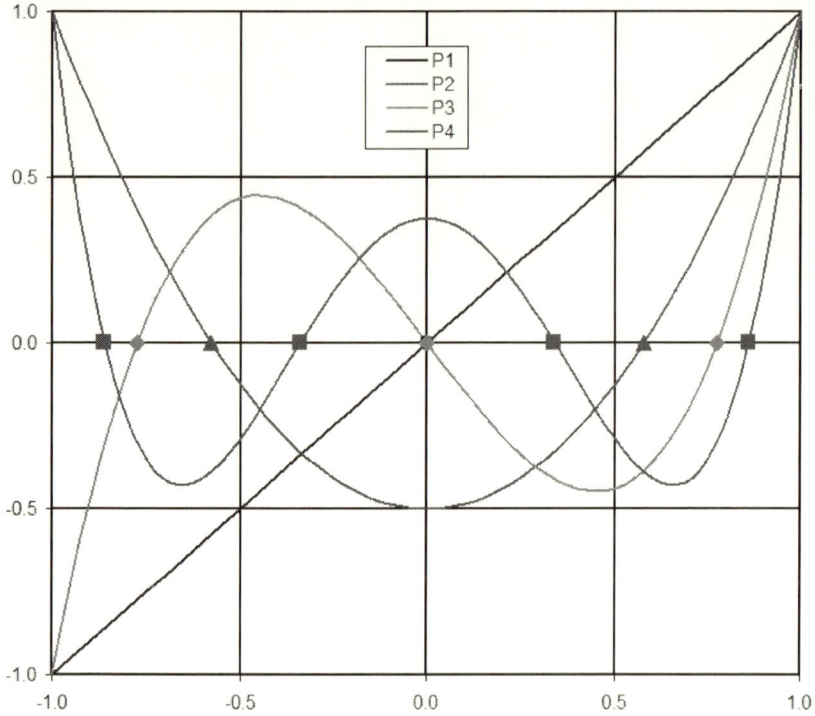

The same procedure with minor modifications will suffice and is included in the Excel® spreadsheet Legendre.xls, which can be found in the online archive in the examples\Legendre folder, along with the complete C code to find any number of roots more accurately using the Newton-Raphson Method. The function calculates the polynomial and its derivative to facilitate this process in a minimal number of steps:

```
void Legendre(int n,double X,double*P3,double*Q3)
  {
  int i;
  double P1,P2,Q1,Q2;
  *P3=1;
  *Q3=0;
  if(n<1)
     return;
```

```
      P2=*P3;
      Q2=*Q3;
      *P3=X;
      *Q3=1;
      if(n<2)
         return;
      for(i=2;i<=n;i++)
         {
         P1=P2;
         Q1=Q2;
         P2=*P3;
         Q2=*Q3;
         *P3=2*X*P2-P1-(X*P2-P1)/i;
         *Q3=Q1+(2*i-1)*P2;
         }
      }
```

The roots are estimated, refined, and inserted into a list in ascending order by the following code:

```
   double RefineRoot(int n,double X1,double X2)
      {
      int i;
      double dP,P,X,Xo;
      X=(X1+X2)/2;
      for(i=0;i<20;i++)
         {
         Legendre(n,X,&P,&dP);
         if(fabs(dP)<0.5)
            break;
         Xo=X;
         X=max(X1,min(X2,X-P/dP));
         if(fabs(P)<=tiny||fabs(X-Xo)<=tiny)
            return(X);
         }
      return(2);
      }

   int InsertRoot(double*Roots,int nr,double Root)
      {
      int i,j;
      for(i=0;i<nr;i++)
         if(fabs(Roots[i]-Root)<tiny)
            return(0);
      for(i=1;i<nr;i++)
         if(Roots[i-1]<Root&&Root<Roots[i])
            break;
      for(j=nr;j>i;j--)
         Roots[j]=Roots[j-1];
      Roots[i]=Root;
```

```c
    return(i);
    }

int FindRoots(int n,double*Roots,double*Weights)
  {
  int i,j,k,nr;
  double dP,dX,P,Root,X,X1,X2;
  printf("computing coefficients\n");
  if(n<2)
    return(0);
  nr=1;
  if(n%2)
    Roots[0]=0;
  else
    {
    Roots[0]=RefineRoot(n,0.,1.3/pow(n,0.94));
    if(n<3)
      goto weights;
    }
  Roots[nr++]=RefineRoot(n,1.-2.1/pow(n,1.9),1);
  if(n<5)
    goto weights;
  printf("\r%i roots found out of %i",nr,(n+1)/2);
  while(nr<(n+1)/2)
    {
    for(i=1;i<nr;i++)
       {
       k=(n+1)/2-nr+1;
       dX=(Roots[i]-Roots[i-1])/k;
       X2=Roots[i-1];
       for(j=0;j<k;j++)
          {
          X1=X2;
          X2+=dX;
          Root=RefineRoot(n,X1,X2);
          if(Root>1.5)
             continue;
          if(!InsertRoot(Roots,nr,Root))
             continue;
          nr++;
          printf("\r%i roots found out of %i",nr,(n+1)/2);
          break;
          }
       if(nr>=(n+1)/2)
          goto weights;
       }
    }
  weights:
  if(n>4)
```

```
    printf("\n");
for(i=0;i<(n+1)/2;i++)
   {
   X=Roots[i];
   Legendre(n,X,&P,&dP);
   Weights[i]=2/(1-X*X)/dP/dP;
   }
return((n+1)/2);
}
```

Chapter 3. Functions of Several Variables

We will next consider problems, which have more than one variable. Most often, the desired solution is a global minimum or maximum. Minima and/or maxima are called extrema (singular: extremum). This process might be illustrated by finding the bottom of the following surface:

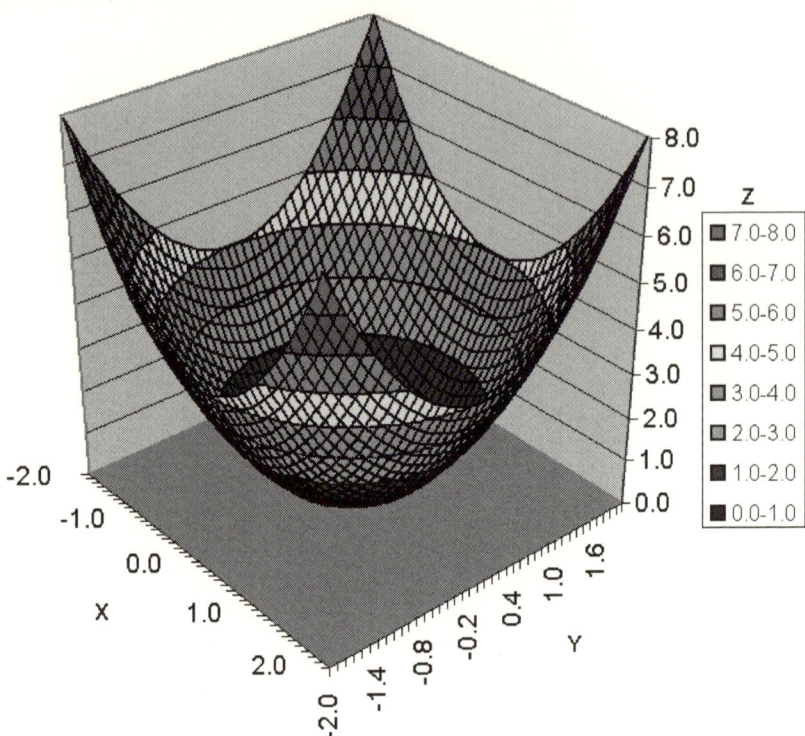

This surface is defined by $z=x^2+y^2$. The low occurs at $x=y=0$ where $\partial z/\partial x = \partial z/\partial y = 0$. Extending Equation 1.1 to multiple dimensions, we arrive at:

$$\begin{bmatrix} x_{k+1} \\ y_{k+1} \end{bmatrix} = \begin{bmatrix} x_k \\ y_k \end{bmatrix} - \frac{1}{damp} \begin{bmatrix} \frac{\partial^2 z}{\partial x^2} & \frac{\partial^2 z}{\partial x \partial y} \\ \frac{\partial^2 z}{\partial x \partial y} & \frac{\partial^2 z}{\partial y^2} \end{bmatrix}^{-1} \begin{bmatrix} \frac{\partial z}{\partial x} \\ \frac{\partial z}{\partial y} \end{bmatrix} \quad (3.1)$$

The square matrix containing the second partial derivatives (or it's inverse) is called a Hessian. For this first problem (first tab in bump.xls) the partial derivatives are easily calculated analytically:

$$\frac{\partial z}{\partial x} = 2x$$
$$\frac{\partial z}{\partial y} = 2y$$
$$\frac{\partial^2 z}{\partial x^2} = \frac{\partial^2 z}{\partial y^2} = 2 \qquad (3.2)$$
$$\frac{\partial^2 z}{\partial x \partial y} = 0$$

The iteration described in Equation 3.1 can be programmed in Excel® using the MMINVERSE(), MMULT(), and TRANSPOSE() functions. In this case, it looks like:

	starting at (0.5,0.5)
x	y	dz/dx	dz/dy	d²z		step
0.50	0.50	1.00	1.00	2.00	0.00	0.500
				0.00	2.00	0.500
0.00	0.00	0.00	0.00			
	starting at (-0.5,0.5)
x	y	dz/dx	dz/dy	d²z		step
-0.50	0.50	-1.00	1.00	2.00	0.00	-0.500
				0.00	2.00	0.500
0.00	0.00	0.00	0.00			
	starting at (0.5,-0.5)
x	y	dz/dx	dz/dy	d²z		step
0.50	-0.50	1.00	-1.00	2.00	0.00	0.500
				0.00	2.00	-0.500
0.00	0.00	0.00	0.00			
	starting at (-0.5,-0.5)
x	y	dz/dx	dz/dy	d²z		step
-0.50	-0.50	-1.00	-1.00	2.00	0.00	-0.500
				0.00	2.00	-0.500
0.00	0.00	0.00	0.00			

Because the first derivatives are linear and the second are constant, the process converges in a single step, regardless of the starting value. Such is not the case with any meaningful problem. We will next consider the function $z=9x/exp(x^2+y^2)$, which has both a minimum and a maximum. The iteration for this problem also converges quickly, as long as a fortuitous initial guess is provided. Whether the iterations end up at the low or high spot depends on the starting value, much as would be the case if you placed a marble on the surface and let it roll into the valley or released a little balloon beneath the surface and let it rise up into the peak. The surface looks like:

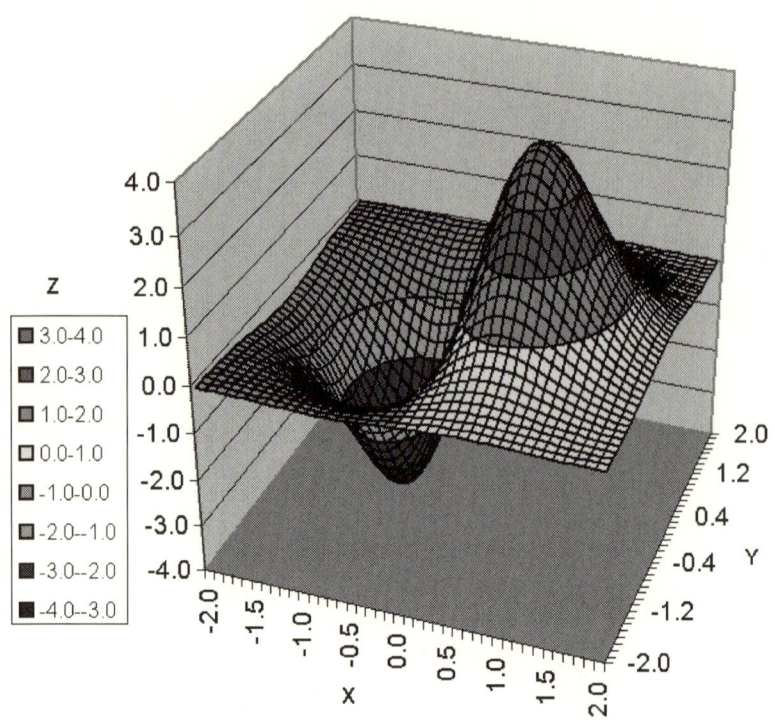

Starting at x=0.5, y=0.5, the iterations proceed as follows:

start at (0.5,0.5) looking for maximum						
x	y	dz/dx	dz/dy	d²z		step
0.50	0.50	2.73	-2.73	-13.65	-2.73	-0.50
		z=	2.73	-2.73	-2.73	1.50
0.75	-0.25	-0.60	1.81	-13.55	-0.30	0.05
		z=	3.61	-0.30	-6.32	-0.29
0.72	-0.11	-0.26	0.81	-14.88	-0.06	0.02
		z=	3.81	-0.06	-7.46	-0.11
0.72	-0.05	-0.13	0.40	-15.21	-0.01	0.01
		z=	3.85	-0.01	-7.66	-0.05
0.71	-0.03	-0.06	0.20	-15.34	0.00	0.00
		z=	3.86	0.00	-7.70	-0.03
0.71	-0.01	-0.03	0.10	-15.39	0.00	0.00
		z=	3.86	0.00	-7.72	-0.01
0.71	-0.01	-0.02	0.05	-15.42	0.00	0.00
		z=	3.86	0.00	-7.72	-0.01
0.71	0.00	-0.01	0.02	-15.43	0.00	0.00
		z=	3.86	0.00	-7.72	0.00
0.71	0.00	0.00	0.01	-15.43	0.00	0.00
		z=	3.86	0.00	-7.72	0.00

After 8 steps, the solution arrives at x=0.71, y=0.0, z=3.86, which is the top of the bump. The Hessian is outlined in red, the right side of Equation 3.1 is outlined in blue, and the left side (i.e., the next step) is outlined in green. If the iteration begins closer to the pit, it will converge to the minimum.

\multicolumn{6}{c	}{start at (-0.5,-0.5) looking for minimum}
x	y	dz/dx	dz/dy	d²z	step
-0.50	-0.50	2.73	-2.73	13.65 2.73	0.50
		z=	-2.73	2.73 2.73	-1.50
-0.75	0.25	-0.60	1.81	13.55 0.30	-0.05
		z=	-3.61	0.30 6.32	0.29
-0.72	0.11	-0.26	0.81	14.88 0.06	-0.02
		z=	-3.81	0.06 7.46	0.11
-0.72	0.05	-0.13	0.40	15.21 0.01	-0.01
		z=	-3.85	0.01 7.66	0.05
-0.71	0.03	-0.06	0.20	15.34 0.00	0.00
		z=	-3.86	0.00 7.70	0.03
-0.71	0.01	-0.03	0.10	15.39 0.00	0.00
		z=	-3.86	0.00 7.72	0.01
-0.71	0.01	-0.02	0.05	15.42 0.00	0.00
		z=	-3.86	0.00 7.72	0.01
-0.71	0.00	-0.01	0.02	15.43 0.00	0.00
		z=	-3.86	0.00 7.72	0.00
-0.71	0.00	0.00	0.01	15.43 0.00	0.00
		z=	-3.86	0.00 7.72	0.00

After 8 steps, the solution arrives at x=-0.71, y=0, z=-3.86, the bottom of the pit. Following the direction defined by the inverse of the Hessian matrix times the first derivatives has often been called the *Method of Steepest Descent* (or *Gradient Descent*). These two examples illustrate this principle quite well. There are two problems with this method, which we will now discuss: 1) the method will sometimes overshoot (as in the one-dimensional cases already presented) and 2) more often than not, we don't have neat analytical expressions for the derivatives.

The problem with step length and starting values is much worse that this illustration would seem to indicate. In fact, I chose the starting values (-0.5,-0.5) and damping factor (*damp* in Equation 3.1 equals 2) so that a solution would be found. There are countless other choices that—rather than improving—get worse with each iteration. Consider the case of starting at x=y=0.6 with no damping (*damp*=1). The first step hops over the solution and almost out of the graph altogether. The second and subsequent steps are worse still. These are illustrated in this next figure:

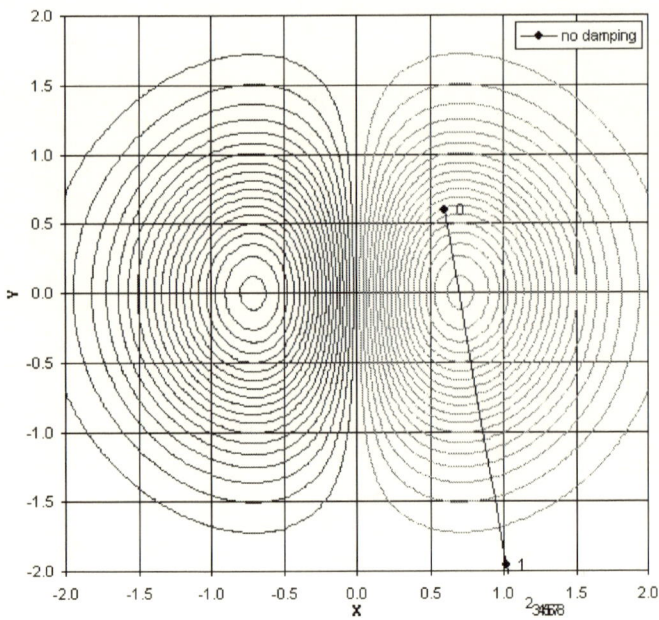

A damping factor of 1.6 is hardly better:

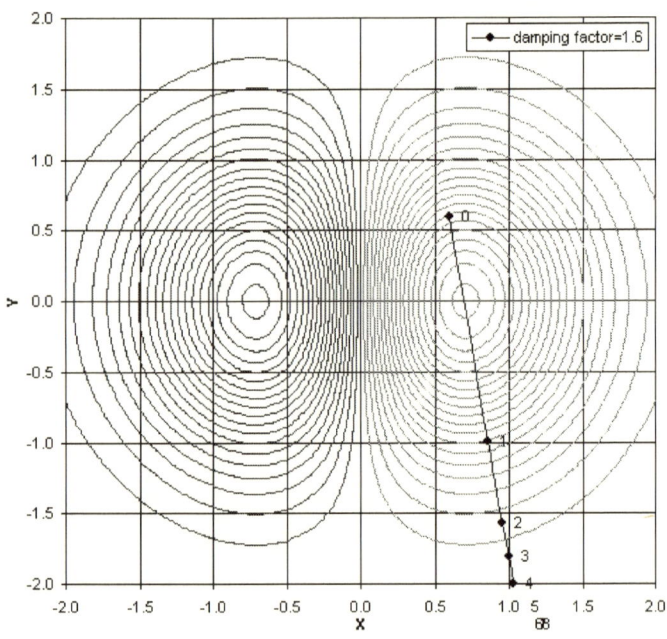

Starting at x=y=1 and *damp*=5 takes off in the wrong direction:

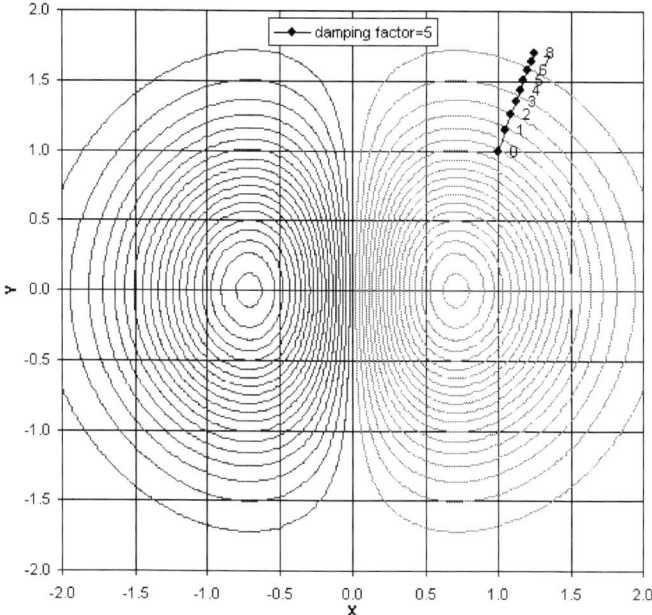

At least starting with x=y=0.5 and *damp*=5 results in a solution.

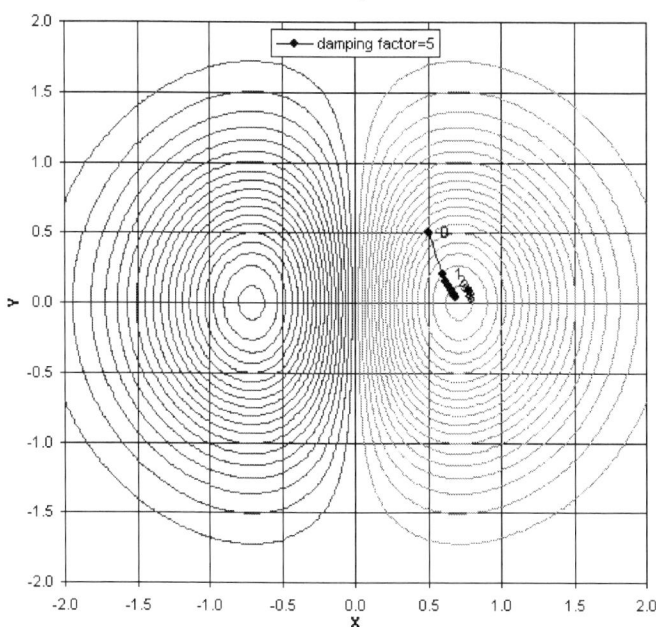

If the method doesn't work reliably or even predictably for this simple case, how can we solve practical problems? The third example in this last set of four shows that the iterations may even take off in the wrong direction. Heavily damping (i.e., a large damping factor) and lots of small iterations is neither efficient nor sufficient in many cases. The figure below shows the starting values (initial guess) that ultimately converge using Newton's Method with a damping factor of 5 for the preceding problem.

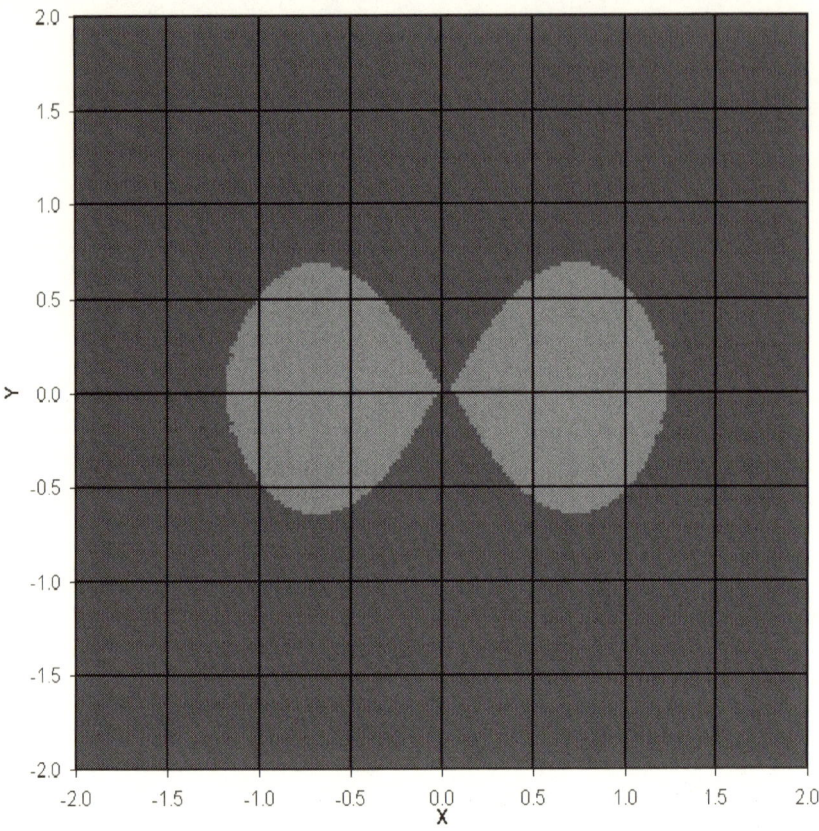

The green regions are sometimes called *attraction* zones. In this case, these zones are quite small, considering the range of possible values. Even within these zones it takes on average 32 iterations to converge. It is no wonder that considerable effort has been devoted to solving these types of problems and various methods abound.

Newton methods take the following form:

$$\begin{bmatrix} next \\ step \end{bmatrix} = \begin{bmatrix} this \\ step \end{bmatrix} - \alpha \begin{bmatrix} Hessian \\ of \\ function \end{bmatrix} \begin{bmatrix} gradient \\ of \\ function \end{bmatrix} \quad (3.3)$$

where α is the step length parameter. We don't always have the second partial derivatives, although the first derivatives are often available; thus, the gradient can be easily calculated, but the Hessian cannot. As the preceding examples illustrate, the simple Newton Method may be inadequate.

Recall that the iterations in the third case went off in the opposite direction of the minimum? If the solution doesn't improve, why not try reversing the step? In the first two cases, the iterations overshot the solution, so why not try 1/2, 1/4, 1/8, etc. A simple modification that often greatly improves convergence is the following sequence of damping factors: 1, -1, 2, -2, 4, -4, 8, -8, ..., which is produced by the following code for two variables, which is easily extended to an arbitrary dimension:

```
r1=residual(x1,y1);
damp=1.;
for(j=0;j<16;j++)
  {
  x2=x1-drdx/damp;
  y2=y1-drdy/damp;
  r2=residual(x2,y2);
  if((increasing&&r2>r1)||((!increasing)&&r2<r1))
    break;
  if(j%2)
   damp=-damp*2.;
  else
    damp=-damp;
  }
```

Chapter 4: Bounding & Scaling

More often than not, valid solutions are limited to some range of values. It is rare that any value (i.e., $-\infty<x<\infty$) will do for a solution. Many of the methods proposed for solving nonlinear problems ignore this important fact. In most cases, this easily handled by the following transformation:

$$x = x_{min} + (x_{max} - x_{min})\left[\frac{1+\tanh(x')}{2}\right] \quad (4.1)$$

Regardless of the value of x', x will be limited to $x_{min}<x<x_{max}$. Near the center, $(x_{max}+x_{min})/2$, the relationship is approximately linear. The hyperbolic tangent function is available in Excel® as well as most programming languages, including C. Most of the range is included between -2 and +2, so that the figures in the preceding chapter serve as useful illustrations of this domain. Another transformation often suggested is:

$$x' = \frac{ax}{(x-x_{min})(x_{max}-x)} \quad (4.2)$$

both are shown in the following figure:

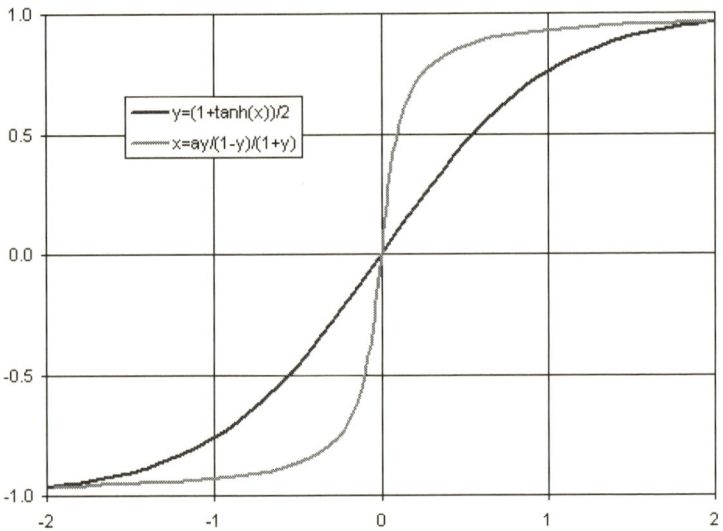

For most applications, Equation 4.2 is much too steep and results in unwanted oscillations. It is also more difficult to implement, involving branches and the square root. Some have suggested using $sin(x)$ or $cos(x)$ instead of $tanh(x)$ in Equation 4.1, but this is not a good choice. It is best to stick with something that has one and only one value of x' for any value of x.

Scaling can be important too. The simplest scaling would be to divide the residual by the span (maximum minus minimum values) or the sum of the squares of the target values. It is rare that a nonlinear problem has an exact solution (i.e., one resulting in a zero residual or precise agreement with the objective). More often than not, nonlinear problems consist of minimizing or maximizing some residual. Many algorithms are designed to either minimize or maximize, but not both. As long as the residual is never zero, the reciprocal of one is the other.

Most any method will work best when a map of the residual looks like a salad bowl, such as the first figure in Chapter 3. Drop a marble anywhere inside a salad bowl and it will eventually find the bottom. Any method worth publishing will converge under the right conditions. It's equally true that most methods will diverge under the wrong conditions. Regardless of how many dimensions (i.e., independent variables) you're working with, it's wise to formulae the problem statement so that the residual approximates this shape.

Residuals may vary over orders of magnitude. When this is the case, you will likely have trouble finding a suitable solution. As the residual is rarely zero, taking the log of the residual may greatly improve the search. Taking the log may also complicate refinement of the solution, as slight variations in the residual may disappear. When this happens, use the log to find an approximate solution, then refine the solution without the log.

Chapter 5: Nonlinear Least-Squares

One of the most common tasks requiring solution of nonlinear equations is that of curve fitting when the function (or functions) being fit to the data are not linear. Sometimes this arises from the shape of the data and sometimes it arises from knowing the analytical solution but not the properties. For example, data, which exhibits one or more asymptotes (i.e., goes straight up or down) or levels off to some constant value. Polynomials (i.e., $y=a+bx+cx^2+...$) will never exhibit this behavior, regardless of how many terms you add. Data that goes straight up or down may sometimes be approximated by: $y=a/(x-b)$. Asymptotic behavior may sometimes be approximated by: $y=(a+bx)/(1+cx)$. Solving this problem so as to minimize the sum of the squares of the residuals (i.e., the difference between the approximation and the data) is called *nonlinear least squares*.

The first problem we will consider here is one, which arises from a known analytical solution. The response begins at $t=f=0$, rises to a maximum, and then falls off asymptotically. The functional shape is $f(t)=at/e^{-t/\tau}$. The data and approximation are shown in the following figure:

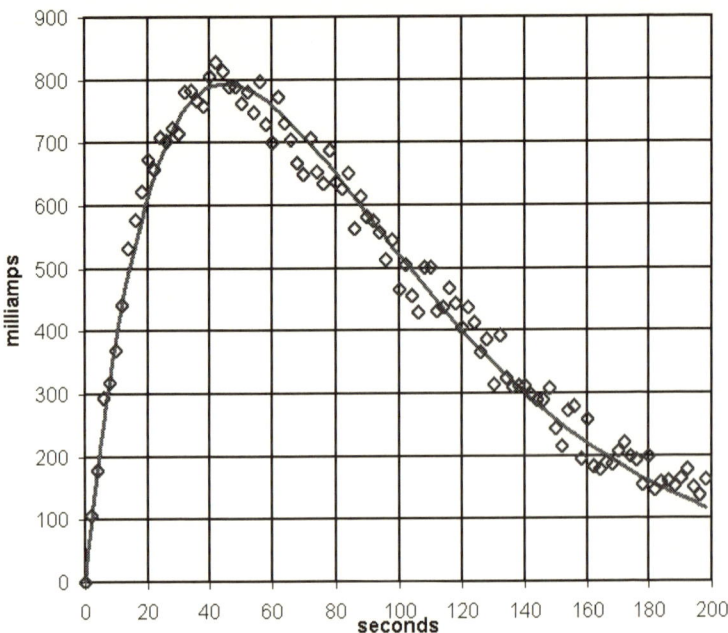

The parameters a and τ are easily calculated with Excel® using the Solver Add-In. Simply guess initial values, create a column of calculated values, then a column of error, and finally set a cell to the sum of the squares of the errors. The Solver will adjust the two (or more) parameters to minimize SUMSQ(). The result yielding the red curve is: a=47.8, τ=45.2. This example can be found in the online archive in folder examples\nllsq nonlinear_least_squares.xls.

The Excel® Solver Add-In is a remarkable tool, which uses several algorithms to obtain a solution. Often this is adequate, but we want to explore how this works, why it sometimes doesn't work, and how to automate and customize the technique in the event we need to process data in production mode or inside a larger software application. The log of the residual for this problem is shown below:

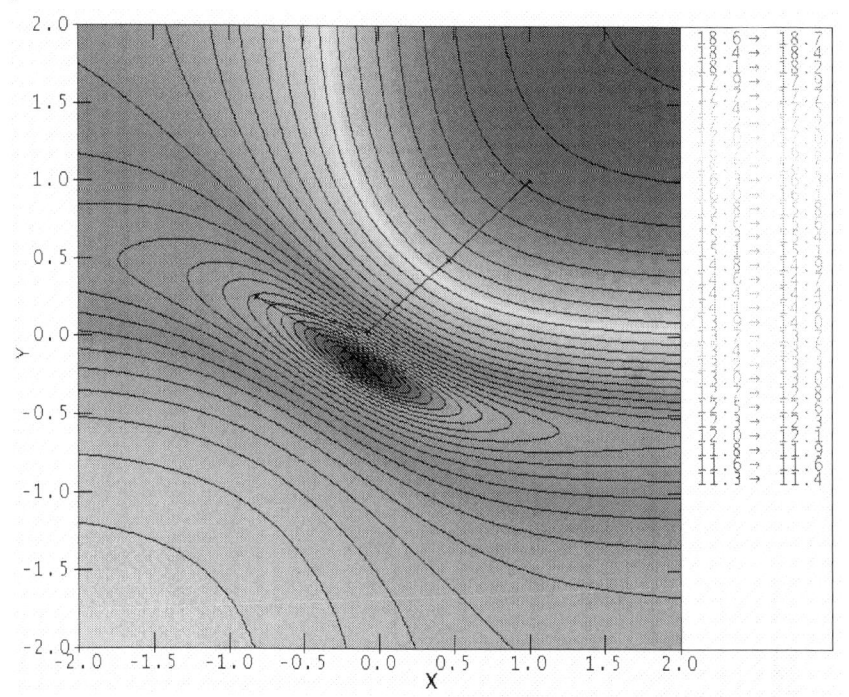

The minimum and maximum values of both a and τ are 25 and 75, respectively. The transform coordinates x and y are used with Equation 4.1 to span these values so that the iterations can't possibly get too far out of bounds. The optimum solution is at the center of the innermost blue ring. Newton's Method (presented in Chapter 3 along with the modification at the end of the chapter) will quickly find a solution, as illustrated by the segmented path above. The code can be found in nllsq.c.

A similar problem also involves inferring properties from experimental data. In this case, the response is linear at first, rises to a maximum, and levels off at that value. The ultimate value is easily calculated by average the final portion of the data. Determining the time constant, τ, and shape factor, a, is a problem in nonlinear minimization. The analytical function that best exhibits this behavior is the hyperbolic tangent, $y=c*tanh((t/tau)^a)$.

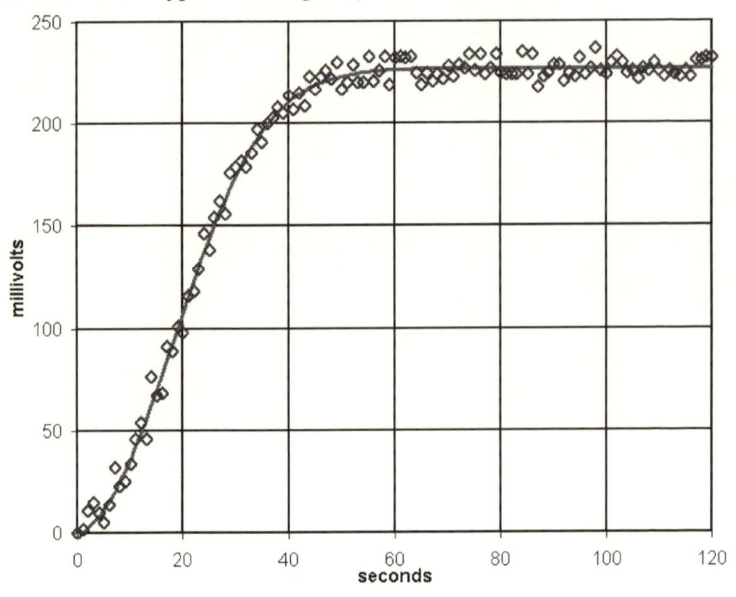

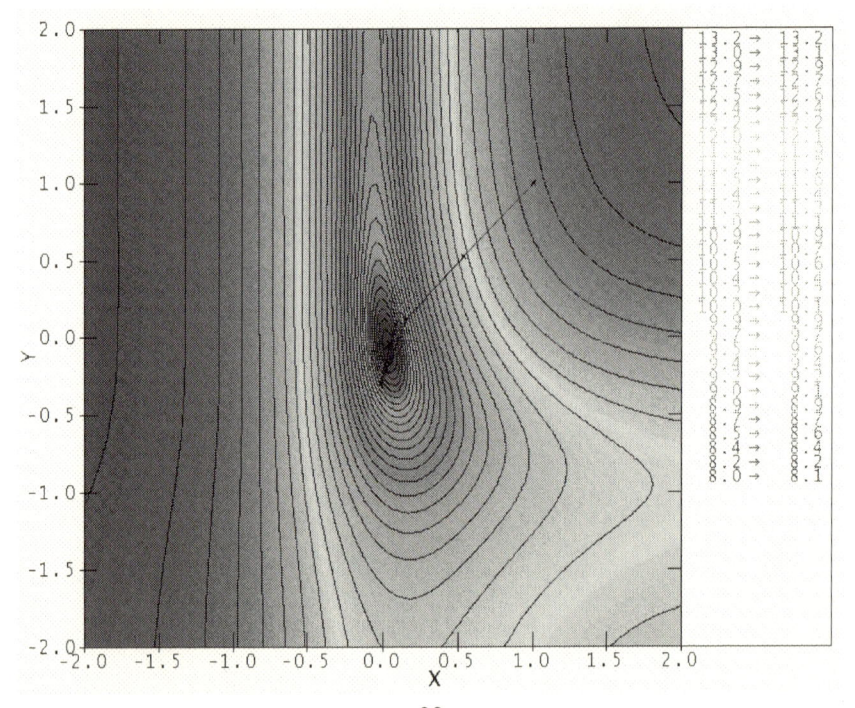

This problem can also easily be solved using the Excel® Solver as before. Supply initial guesses, create a column of calculated values, a column of errors, and a cell containing SUMSQ(). The asymptotic value, c=227, the time constant, τ=29.8, and the exponent is 2/3. The data and solution are shown in the preceding figure. It is not necessary (or helpful) to take the log of the residual in this case. The residual and solution path is mapped below in the transform coordinates.

In both of these examples, the gradient and Hessian are calculated using finite differences. This is often inconvenient, slow, and inaccurate. Such is often the case, providing more than adequate motivation for developing what are called *Quasi-Newton Methods*, that is, methods that are roughly based on Newton's Method but do not require the gradient and/or Hessian.

Chapter 6: Quasi-Newton Methods

The need to solve practical problems has given rise to what are called *Quasi-Newton Methods*, that is, methods that are roughly based on Newton's Method, but have been modified to address the lack of convenient analytical expressions for the partial derivatives.

Davidon-Fletcher-Powell Method

DFP was the first technique to extend the Secant Method to multi-dimensions.[6] DFP is a rank-two method that uses the following update formula for the inverse of Hessian matrix, $B=H^{-1}$:

$$B_{k+1} = B_k + \gamma s_k s_k^T + \beta p_k p_k^T \qquad (6.1)$$

$$s_k = x_{k+1} - x_k \qquad (6.2)$$

$$y_k = \nabla f(x_{k+1}) - \nabla f(x_k) \qquad (6.3)$$

$$p_k = B_k y_k \qquad (6.4)$$

$$\gamma = \frac{1}{y_k^T s_k} \qquad (6.5)$$

$$\beta = -\frac{1}{y_k^T p_k} \qquad (6.6)$$

Scalars γ and β plus the vector p are selected so as to follow the Secant Method and achieve second order agreement with a Taylor series expansion of the function being approximated. This iterative update also assures that the Hessian is symmetric and positive definite. The direction vector, d_k, arises from Equation 3.1, where Δ_k is the gradient at x_k:

$$d_k = -B_k \nabla f(x_k) \qquad (6.7)$$

The iteration typically begins with B equal to the identity matrix. As we have already seen, assuming a positive step length, $\alpha>0$, may lead away from the solution. It may also take quite a few steps to find the value of α that minimizes (or maximizes) the function (i.e., minimizes the gradient along the direction d_k.

---

[6] Johnson, I. L., Jr., "The Davidon-Fletcher-Powell Penalty Function Method: A Generalized Iterative Technique for Solving Parameter Optimization Problems," NASA Technical Note D-8251, 1976.

We will first consider the problem in which $z=(cosh(x-0.5)+(y-0.5)^2)/8$, shown in the following figure:

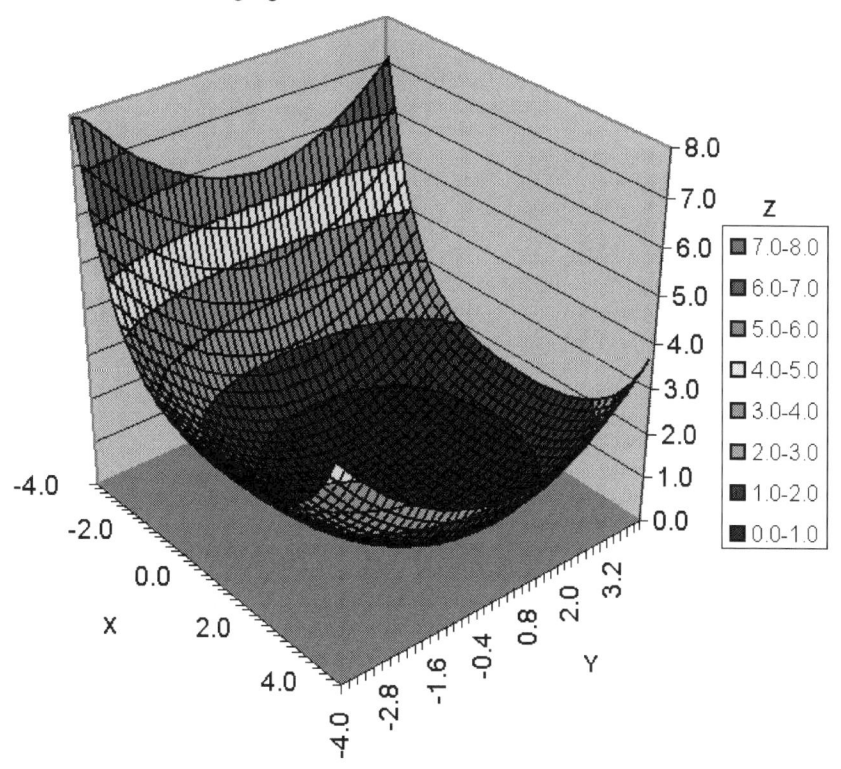

Beginning with a guess of $x=y=-3$, we apply the DFP Method along with Brent's Method to locate the minimum, which is at $x=y=0.5$. The source code (dfpmin.c), equations, and accompanying spreadsheet can be found in the examples\dfp folder. The solution follows the path shown in the following figure. This is a very simple problem and the partial derivatives are easily calculated, which is rarely the case with practical applications.

```
double func(double*p)
    {
    calls.func++;
    return((cosh(p[0]-0.5)+sq(p[1]-0.5))/8.);
    }
void grad(double*p,double*g)
    {
    calls.grad++;
    g[0]=sinh(p[0]-0.5)/8.;
    g[1]=(2.*p[1]-1.)/8.;
    }
```

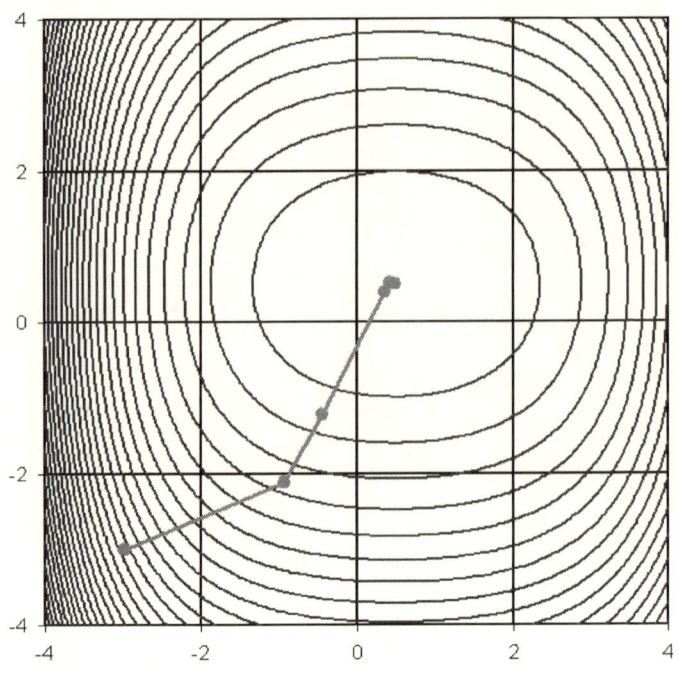

Initializing the Hessian to the identity matrix is often a poor choice. It works in this case because the residual is shaped like a salad bowl and the magnitude of the residual sort of fits with the dimensions of the solution domain. The algorithm itself is quite simple:

```
f2=func(p);
grad(p,g);
for(i=0;i<n;i++)
   {
   for(j=0;j<n;j++)
      h[n*i+j]=0.;
   h[n*i+i]=1.;
   d[i]=-g[i];
   }
for(it=1;it<=itmax;it++)
   {
   a=1.;
   for(jt=0;jt<32;jt++)
      {
      for(i=0;i<n;i++)
         s[i]=p[i]+d[i]/a;
      f2=func(s);
      if(f2<*f)
         break;
      if(jt==0)
```

36

```
      a=2.;
   else if(jt%2)
      a=-a;
   else
      a=-a*2.;
   }
   for(i=0;i<n;i++)
   {
     s[i]=d[i]/a;
     p[i]+=s[i];
   }
   f1=*f;
   *f=f2;
   if(2.*fabs(f1-f2)<=epsilon*(fabs(f1)
   +fabs(f2)+epsilon))
     break;
   for(i=0;i<n;i++)
     y[i]=g[i];
   grad(p,g);
   for(i=0;i<n;i++)
     y[i]=g[i]-y[i];
   for(i=0;i<n;i++)
     for(q[i]=j=0;j<n;j++)
       q[i]+=h[n*i+j]*y[j];
   for(yTs=yHy=i=0;i<n;i++)
   {
     yTs+=y[i]*s[i];
     yHy+=y[i]*q[i];
   }
   for(i=0;i<n;i++)
     for(j=0;j<n;j++)
       h[n*i+j]+=-q[i]*q[j]/yHy+s[i]*s[j]/yTs;
   for(i=0;i<n;i++)
     for(d[i]=j=0;j<n;j++)
       d[i]-=h[n*i+j]*g[j];
}
```

We can also use the DFP Method to solve the two problems from the end of the last chapter. The DFP method does find the correct solution to the first problem, but arrives there through a circuitous path, as shown in the following figure. The gradient is calculated as before by finite difference, but that's not the problem. The poor initial estimate of the Hessian is.

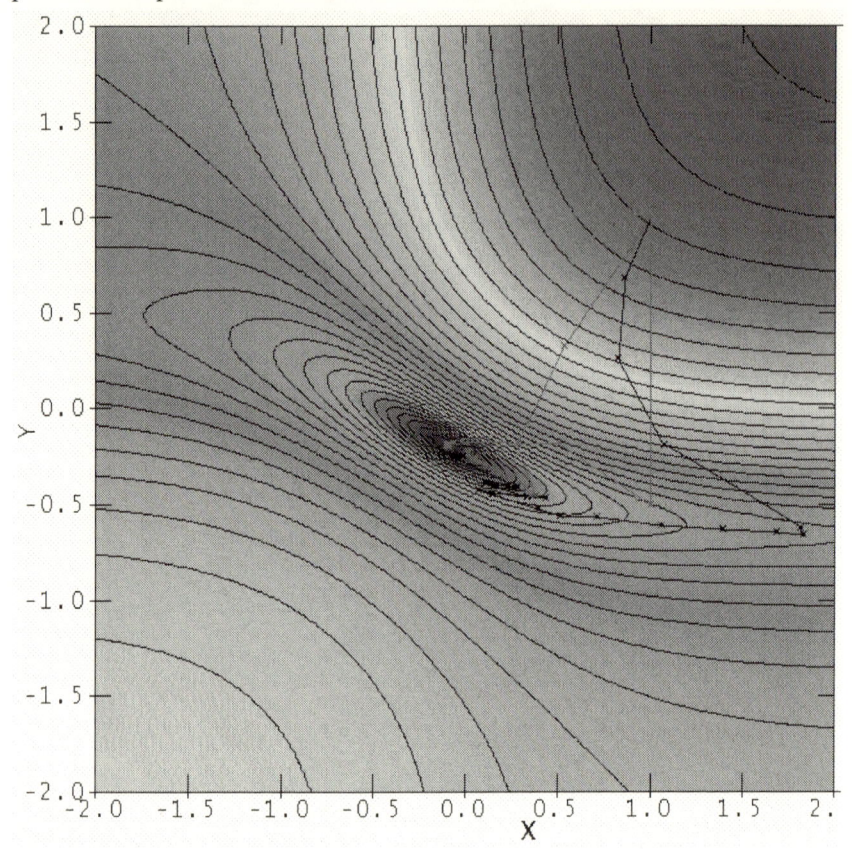

The DFP solution path is shown in black. The magenta line is the solution path for the next method presented below and the dark cyan line is the solution path for the method presented in the next chapter.

A solution is found for the second by a much more direct path, well approximating the steepest descent or Newton's Method. This is purely coincidental.

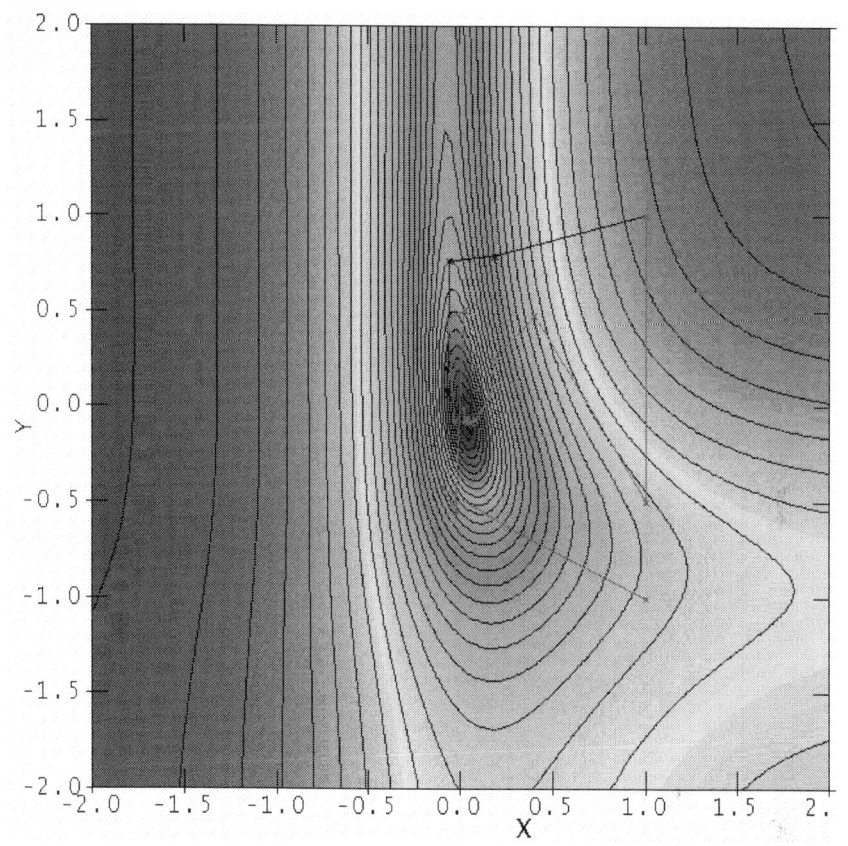

The DFP solution path is shown in black. The magenta line is the solution path for the BGFS method presented below and the dark cyan line is the solution path for Broyden's Method, which is presented in the next chapter.

Broyden-Fletcher-Goldfarb-Shanno Method

The BFGS Method was introduced after the DFP Method and is often said to be preferable to it.[7,8] In practice which method is faster and/or more accurate depends on the problem, as illustrated in the two preceding figures. The BFGS uses a different Hessian update formula:

$$H_{k+1} = H_k + \gamma y_k y_k^T + \beta p_k p_k^T \tag{6.8}$$

$$p_k = H_k s_k \tag{6.9}$$

$$\gamma = \frac{1}{y_k^T s_k} \tag{6.10}$$

$$\beta = -\frac{1}{s_k^T p_k} \tag{6.11}$$

The search direction is given by:

$$H_k d_k = -\nabla f(x_k) \tag{6.12}$$

A variant of the BFGS algorithm that generates the inverse Hessian ($B=H^{-1}$) with each step is given by the following:

$$B_{k+1} = B_k - \gamma\left(q_k s_k^T + s_k q_k^T\right) + \varphi s_k s_k^T \tag{6.13}$$

$$q_k = B_k y_k \tag{6.14}$$

$$\phi = \frac{s_k^T y_k + q_k^T y_k}{\left(s_k^T y_k\right)^2} \tag{6.15}$$

Calculating the gradient with finite differences at each step often takes much longer than inverting a small matrix, so that eliminating the step of solving a small set of linear equations (i.e., inverting H to get B) is of little consequence and may take equally as long as Equation 6.13 over Equation 6.8. The BFGS solution path is shown in magenta in the two preceding figures.

---

[7] Goldfarb, D., "A Family of Variable Metric Methods Derived by Variational Means," Mathematics of Computation Journal, Vol. 24, pp. 23–26, 1970.
[8] Shanno, D. F., "Conditioning of Quasi-Newton Methods for Function Minimization," Mathematics of Computation Journal, Vol. 24 pp. 647–657, 1970.

An exploded view in the vicinity of the solution is shown below:

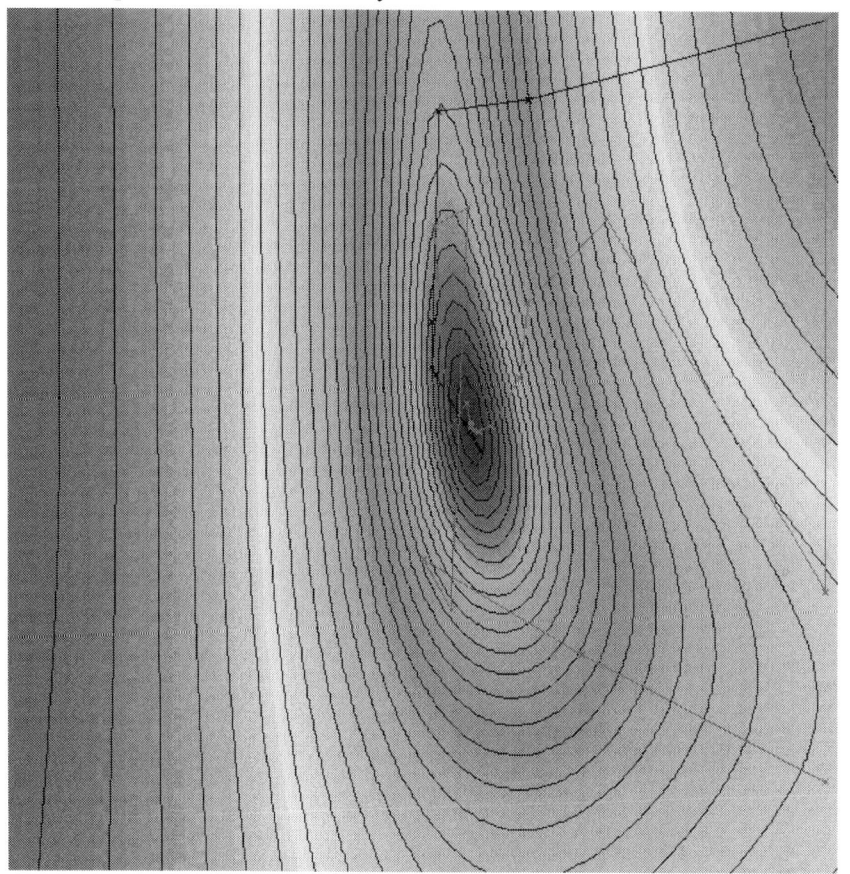

While the DFP and BFGS algorithms are suitable for many problems, it would be *really* convenient to have a method that didn't require derivatives!

Chapter 7. Broyden's Derivative-Free Algorithm

Broyden—one of the developers of the BFGS Method—also devised a derivative-free method.[9] This technique is particularly attractive, as the gradient and Hessian are rarely available for practical problems. In the next chapter we will consider another way of dealing with this problem. We will first consider Broyden's method, which doesn't require derivatives. This method uses all the residuals (instead of just the sum of the squares), which often greatly exceeds the number of independent parameters, especially in regression problems.

Rather than starting with Newton's Method and updating an estimate of the Hessian, Broyden's begins with the *Jacobian*. The elements of the Jacobian are given by:

$$J_{i,j} = \frac{\partial r_i}{\partial x_j} \qquad (7.1)$$

where r_i is the residual term i. For a least-squares problem, the terms of the Hessian are:

$$H_{i,j} = \frac{\partial^2 (r^T r)}{\partial x_i \partial x_j} \qquad (7.2)$$

The Hessian is always square ($n \times n$); whereas, the Jacobian is often rectangular ($m \times n$, $m > n$). It also stands to reason that we retain more information about the residual in the Jacobian than the Hessian, as it has a greater number of rows. Broyden's method in terms of the same variables as before can be expressed by:

$$J_k s_k = y_k \qquad (7.3)$$

$$J_{k+1} = J_k + \frac{(y_k - J_k s_k) s_k^T}{s_k^T s_k} \qquad (7.4)$$

There is also an inverse update to the Broyden method, but this is often problematic and not worth the savings in solving a few simultaneous linear equations.[10]

$$J_{k+1}^{-1} = J_k^{-1} + \frac{(s_k - J_k^{-1} y_k) y_k^T}{y_k^T y_k} \qquad (7.5)$$

---

[9] Broyden, C., "A New Method of Solving Nonlinear Simultaneous Equations," *Computational Journal*, Vol. 12, pp. 94-99, 1969.
[10] Kvaalen, E., "A Faster Broyden Method," SIAM BIT Numerical Mathematics, Vol. 31, No. 2, pp. 369–372, November, 1991.

The most convenient aspect of Broyden's method is that initialization of the Jacobian is so easily accomplished. Simply hop around the domain a bit and update Equation 7.4. This is a rank-one update and information is only added in the space traversed, eventually approximating the true Jacobian. As for the performance, the Broyden Method solution path has already been illustrated in the two problems of the previous chapter (the dark cyan lines). In both of these examples, the Broyden path is more direct than the DFP, but not quite as direct as the BFGS. In both cases, it takes fewer steps to reach this solution.

I first published a paper on Broyden's Method in 1991, which included some enhancements, including escape from extraneous entrapment in local minima. The text of this paper is included in the final appendix. All of the associated software can be found in the online archive in the folder examples\Broyden. When trapped, simply hop around a bit and it will often escape and find a better solution.

The next problem we will solve using Broyden's Method to is $r_1=75*x_1^2-169*x_2^2/9$, $r_2=845*x_1*x_2^2/3-125*x_1^3-1$. The solution path is shown below:

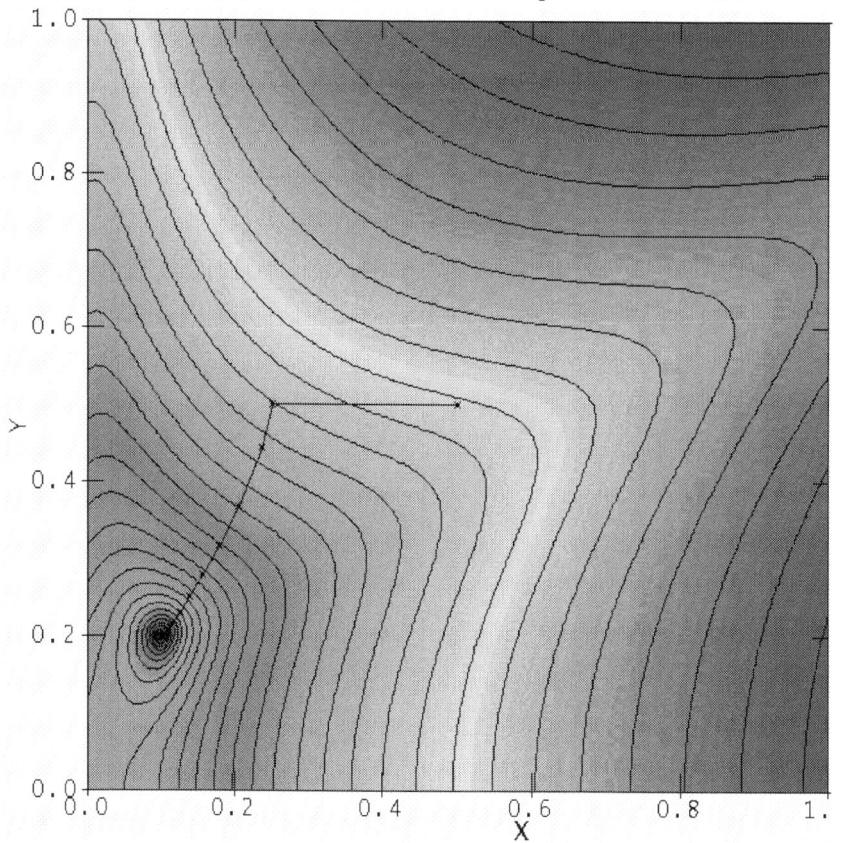

Of course, many problems have more than two unknowns and can't be illustrated in two-dimensional space. The following problem has three: $f_1=15*x_1-\cos(5/9*(x_2-1/5)*\pi*x_3)-1/2$, $f_2=25*x_1^2-81*(x_2-1/10)^2-\sin(5*\pi*x_3/9)+1$, $f_3=\exp(-5*x_1*(x_2-1/5))-100*\pi*x_3/9+10*\pi/3-1$. One way of presenting the solution path is a shaded cube as seen from three sides:

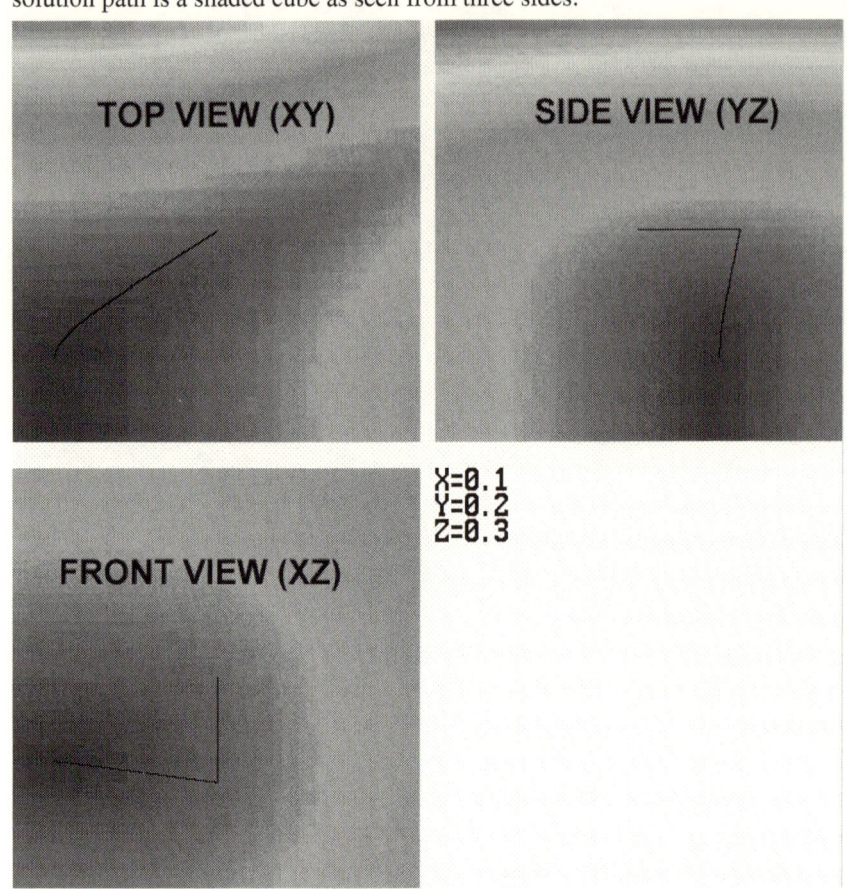

The solution path starts in the middle ($x_1=x_2=x_3=0.5$) and proceeds to the center of the blue spot, near $x_1=0.1$, $x_2=0.2$, $x_3=0.3$.

A similar problem is described by the equations: $f_1 = x_1 + \cos(5*(x_1 - 1/10)*x_2*x_3/3) - 11/10$, $f_2 = |x_1 - 11/10|^{1/4} + x_2/2 + 5*x_3^2/9 - x_3/2 - 1$, $f_3 = -(x_1 - 1/10)^2 - x_2^2/40 + x_2/200 + 10*x_3/3 - 1$.

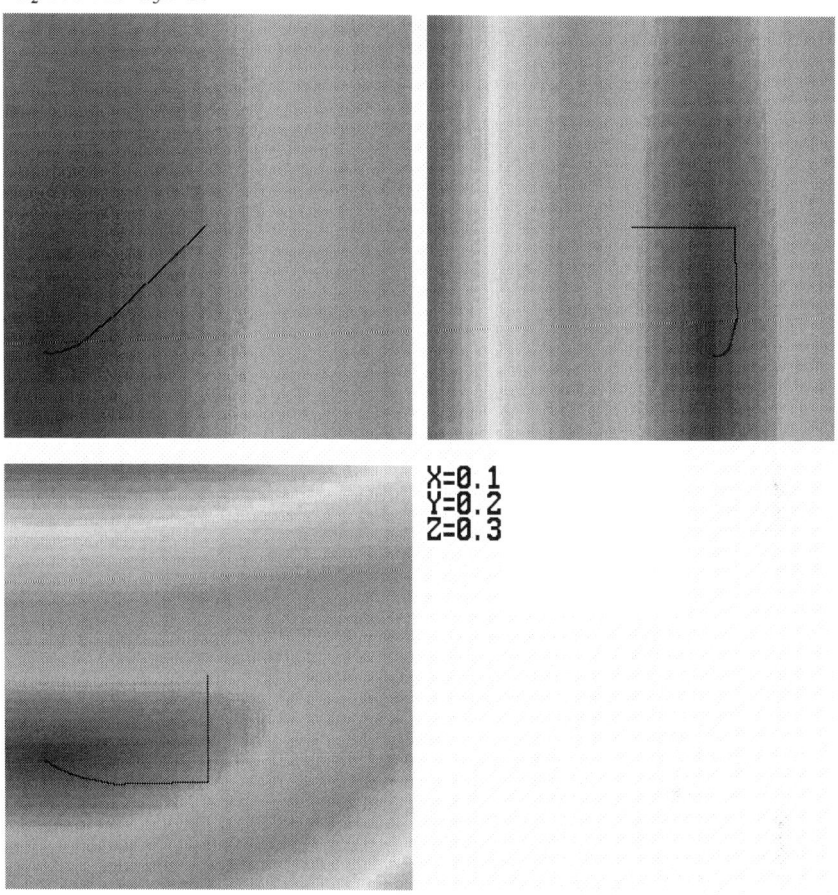

This solution path also starts in the middle ($x_1 = x_2 = x_3 = 0.5$) and proceeds to the center of the blue spot, near $x_1 = 0.1$, $x_2 = 0.2$, $x_3 = 0.3$.

45

Chapter 8. Approximating Gradients and Hessians

In the preceding examples we have already seen the most common way of approximating gradients (first partial derivatives) and Hessians (second partial derivatives)—finite differences. In many cases this will suffice, especially when the residual function is an explicit formula, calculated using double-precision floating-point arithmetic. An entire class of problems exists for which this may not be the case—model-tuning parameters. For instance, we may be calibrating a much more complex model to experimental data. Fields in which such problems arise span the range from ecological systems to the stock market. I have used the following method on many occasions and allude to this in the SIAM paper (Appendix C).

Such problems have one or more of several characteristics, including: 1) results may not be available for any arbitrary inputs, 2) the model may be a separate program that can't be launched or called as function, 3) the number of model runs must be minimized due to computational time or limited resources, 4) the convergence level of the model is such that small changes in tuning parameters produce meaninless differentials.

Case 3 in the SIAM paper is an example of this type problem. The objective was tuning a cooling tower model (FACTS) by simultaneously adjusting heat and mass transfer coefficients. In the original implementation, I created an input file, launched the executable cooling tower model, and read the output file—all inside the nonlinear equation solver loop. It was quite cumbersome and very time consuming. I have modified Case 2 in the online archive to simplify and approximate the program-launching step.

Another program I work with a lot is GateCycle®, a thermodynamic cycle modeling tool sold by General Electric®. GC is especially useful for analyzing combined cycle power systems. While this program is quirky and hasn't been updated since 2005[11], it's still far superior to similar programs currently available. It's fast, but achieves this at the expense of several important behaviors, including convergence. Every time you run GC, you get a slightly different result—even with the exact same inputs. From a programming standpoint, this is completely unacceptable. This problematic behavior arises from the fact that it uses the last solution as the starting point for the next. If the previous solution croaked, you're screwed—another unacceptable outcome. Because the solution always contains some level of noise (imprecise convergence), you can't reliably calculate partial derivatives as you might with a more precise and consistent model.

---

[11] There's a newer version of GateCycle (6.x), but it has the worst user interface of any program ever written—besides being slower than a herd of snails stampeding up the side of a salt dome. GC6 is an xml-based programming atrocity and legendary corporate blunder. General Electric® dumped all of the original developers, who have dispersed to the four winds.

Local Approximation of the Residual

The solution to these challenges is to approximate the residual and it's partial derivatives locally, using a database of model runs. This is how I eventually replaced the cumbersome process in Case 3 of the SIAM paper and how I have tuned many GateCycle® models to historical operational data. The first step is to build a table of runs that span the expected range of tuning parameters. We read the table and use the results to build a least squares approximation of the residual at each step. Since we have the gradient and the Hessian, Newton's Method is the obvious choice. For the case of two tuning parameters (x and y), we have:

$$r = a_0 + a_1 x + a_2 y + a_3 x^2 + a_4 xy + a_5 y^2 \qquad (8.1)$$

Gather enough data points from the table around the current value to assure over-determination of the coefficients—nine in the case of two variables. The code, input tables, and results can be found in the online archive in folder examples\tabular. The approximating function is:

```
double ApproximateSurface(double*p,double*g,double*h)
{
int i,j,x,xx,y,yy;
double A[n],AtA[n*n],AtB[n],B;
for(xx=1;xx<surface.nx-2;xx++)
  if(p[0]<surface.x[xx+1])
    break;
for(yy=1;yy<surface.ny-2;yy++)
  if(p[1]<surface.y[yy+1])
    break;
memset(AtA,0,sizeof(AtA));
memset(AtB,0,sizeof(AtB));
for(y=yy-1;y<=yy+1;y++)
  {
  for(x=xx-1;x<=xx+1;x++)
    {
    B=surface.z[surface.nx*y+x];
    A[0]=1.;
    A[1]=surface.x[x];
    A[2]=surface.y[y];
    A[3]=A[1]*A[1];
    A[4]=A[1]*A[2];
    A[5]=A[2]*A[2];
    for(i=0;i<n;i++)
      {
      for(j=0;j<n;j++)
        AtA[n*i+j]+=A[i]*A[j];
      AtB[i]+=A[i]*B;
      }
    }
```

```
    }
  Gauss(AtA,AtB,n);
  if(g)
    {
    g[0]=AtB[1]+2.*AtB[3]*p[0]+AtB[4]*p[1];
    g[1]=AtB[2]+AtB[4]*p[0]+2.*AtB[5]*p[1];
    }
  if(h)
    {
    h[0]=2.*AtB[3];
    h[1]=h[2]=AtB[4];
    h[3]=2.*AtB[5];
    }
  return(AtB[0]
         +AtB[1]*p[0]
         +AtB[2]*p[1]
         +AtB[3]*p[0]*p[0]
         +AtB[4]*p[0]*p[1]
         +AtB[5]*p[1]*p[1]);
  }
```

The Newton iteration is the same as presented at the end of Chapter 3. Results for the two examples are shown in the following figures:

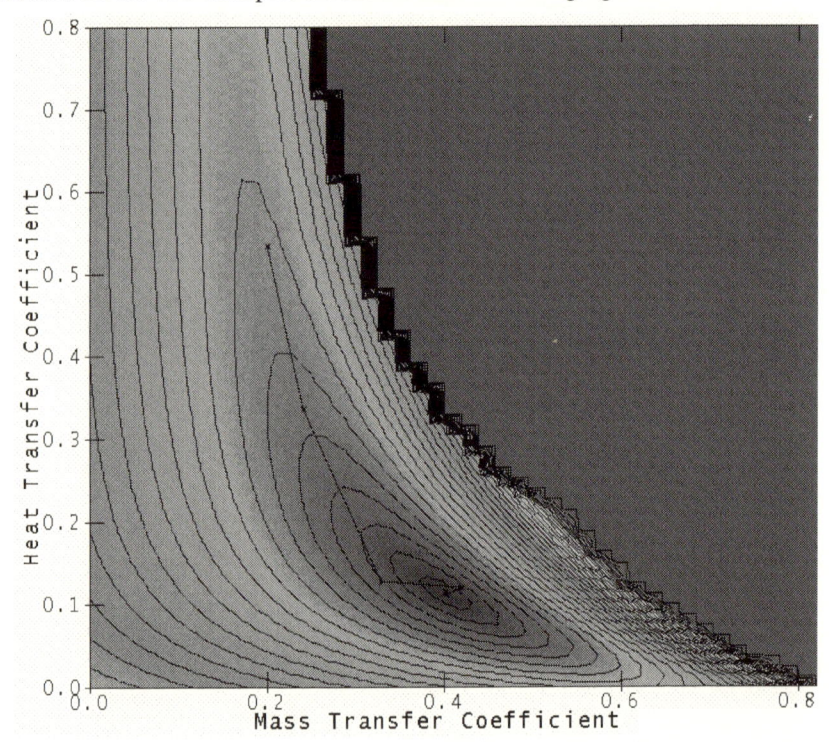

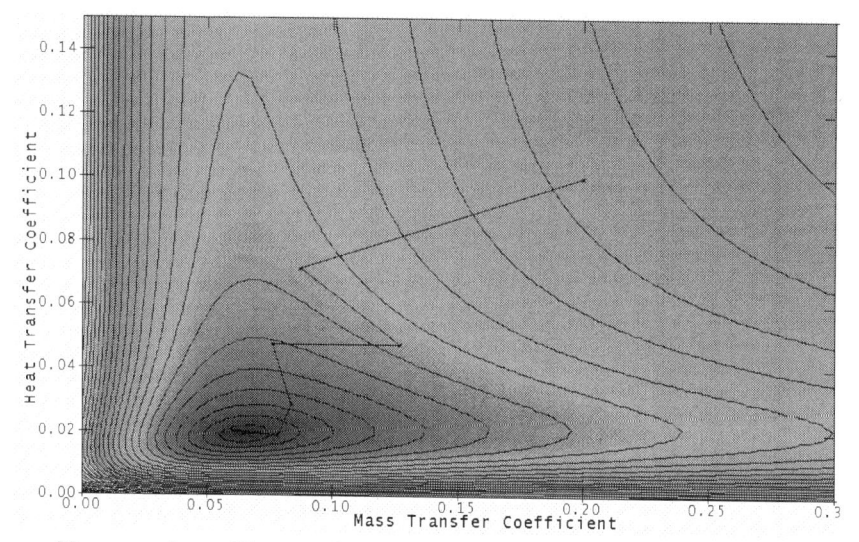

No regression will ever adequately approximate the surfaces illustrated in the preceding two figures. Simply curve-fitting the results and analytically solving for the minimum isn't an option. Merely picking the smallest value in the tabulated results isn't good enough either, considering the coarse resolution.

Broyden's Method from Chapter 7 can also be used with tabulated data, provided the number of residuals is greater than or equal to the number of unknowns. The following example comes directly from a GateCycle® model, in which capacity and heat rate (inversely proportional to thermal efficiency) are calculated based on ambient conditions. The results are shown below:

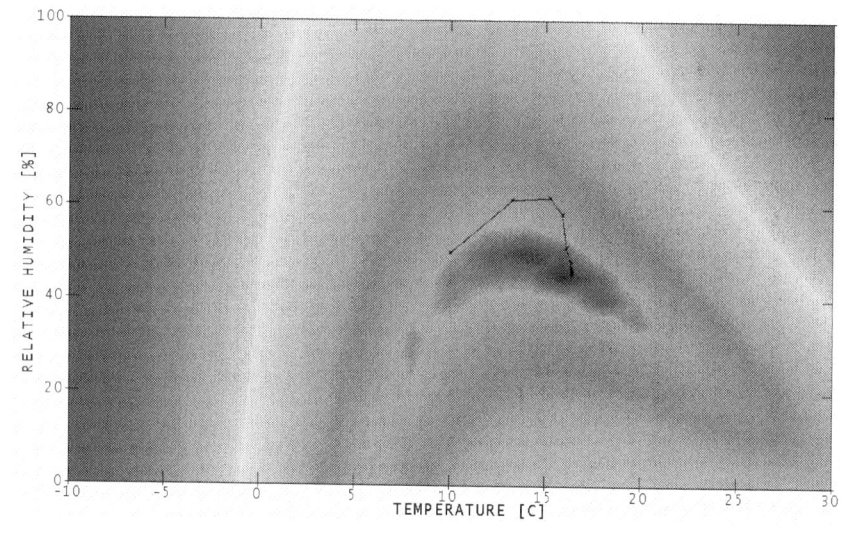

Chapter 9. Multi-Dimensional Bisection Search

In Chapter 1 we considered the number of iterations required to locate a solution and how this may be influenced by the starting location or initial guess. Several of the colored figures revealed large areas where the algorithms failed (Illinois Variant 2 and Anderson-Björck). The last figure in Chapter 3 showed very limited attraction zones (i.e., regions within which an initial guess will result in success of the method, yielding a solution). The first figure in Chapter 8 has a large flat zone (i.e., the whole upper right half of the domain is red). While the gradient vanishes in this zone, it's not the solution we're hoping to find, because it corresponds to a large residual.

Brute Force

The most inefficient way to locate a solution would be to take small steps throughout the entire domain, saving the highest and/or lowest value. For illustration, we will use the second problem from Chapter 2—the "bump" defined by $z=9x/exp(x^2+y^2)$. In order to obtain a solution within $\Delta x=\Delta y=\pm 0.1$, this would require $(4/0.1+1)^2=1681$ function evaluations, as the domain spans -2 to +2 for both x and y. To obtain a solution to within ± 0.01 would require $10^2=100$ times as many function evaluations. If there were three variables instead of two, the former would require 68,921 function evaluations and the latter 69 billion. Almost any intelligent method is more attractive than brute force. Randomly searching a domain makes as much sense as buying lottery tickets. Lotteries are a shameful *Exploitation of the Mathematically Challenged*.

Divide and Conquer

The simplest approach to searching for a solution (or an effective initial estimate from which to further refine a solution) is to divide and conquer. This technique is simple: sequentially subdivide the domain, only refining the search in those subdivisions that yield better (or the best) results. This is easily accomplished with a re-entrant function (i.e., one that calls itself). While this may result in a stack overflow, there's probably more than enough memory on any modern computer to handle it.

The code is simple (example1.c in folder examples\bisection):

```
void bisect(double Xm,double Xx,double Ym,double
    Yx,double func(double,double),int
    n,double*X,double*Y,double*F)
{
double F11,F12,F21,F22,X1,X2,Y1,Y2;
X1=(3.*Xm+Xx)/4.;
X2=(Xm+3.*Xx)/4.;
Y1=(3.*Ym+Yx)/4.;
Y2=(Ym+3.*Yx)/4.;
F11=func(X1,Y1);
F12=func(X1,Y2);
```

```
F21=func(X2,Y1);
F22=func(X2,Y2);
if(F11<=fmin3(F12,F21,F22))
   {
   if(F11<*F)
      {
      *X=X1;
      *Y=Y1;
      *F=F11;
      }
   if(n>0)
      bisect(Xm,(Xm+Xx)/2.,Ym,(Ym+Yx)/2.,func,
   n-1,X,Y,F);
   }
if(F12<=fmin3(F11,F21,F22))
   {
   if(F12<*F)
      {
      *X=X1;
      *Y=Y2;
      *F=F12;
      }
   if(n>0)
      bisect(Xm,(Xm+Xx)/2.,(Ym+Yx)/2.,Yx,func,
   n-1,X,Y,F);
   }
if(F21<=fmin3(F11,F12,F22))
   {
   if(F21<*F)
      {
      *X=X2;
      *Y=Y1;
      *F=F21;
      }
   if(n>0)
      bisect((Xm+Xx)/2.,Xx,Ym,(Ym+Yx)/2.,func,
   n-1,X,Y,F);
   }
if(F22<=fmin3(F11,F12,F21))
   {
   if(F22<*F)
      {
      *X=X2;
      *Y=Y2;
      *F=F22;
      }
   if(n>0)
      bisect((Xm+Xx)/2.,Xx,(Ym+Yx)/2.,Yx,func,
   n-1,X,Y,F);
```

}
}

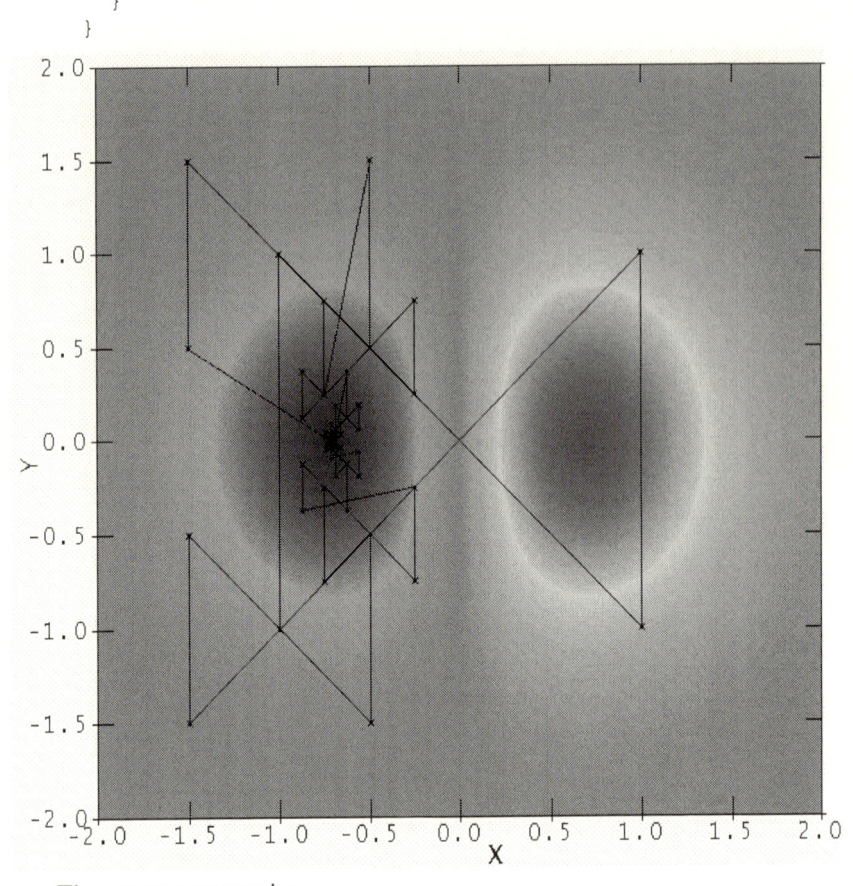

The program output is:

```
searching for solution
n=8, X=-1, Y=-1, F11=-1.21802
n=7, X=-0.5, Y=-0.5, F22=-2.72939
n=6, X=-0.75, Y=-0.25, F12=-3.61301
n=5, X=-0.625, Y=-0.125, F22=-3.74706
n=4, X=-0.6875, Y=-0.0625, F12=-3.84191
n=3, X=-0.71875, Y=-0.03125, F12=-3.85513
n=2, X=-0.703125, Y=-0.015625, F22=-3.85887
n=1, X=-0.710938, Y=-0.0078125, F12=-3.85959
n=0, X=-0.707031, Y=-0.00390625, F22=-3.85988
```

This technique is basically a bisection search in two dimensions. Consider the first example in Chapter 6, $z=(cosh(x-0.5)+(y-0.5)^2)/8$. The solution path for this problem (example2.c) is similar:

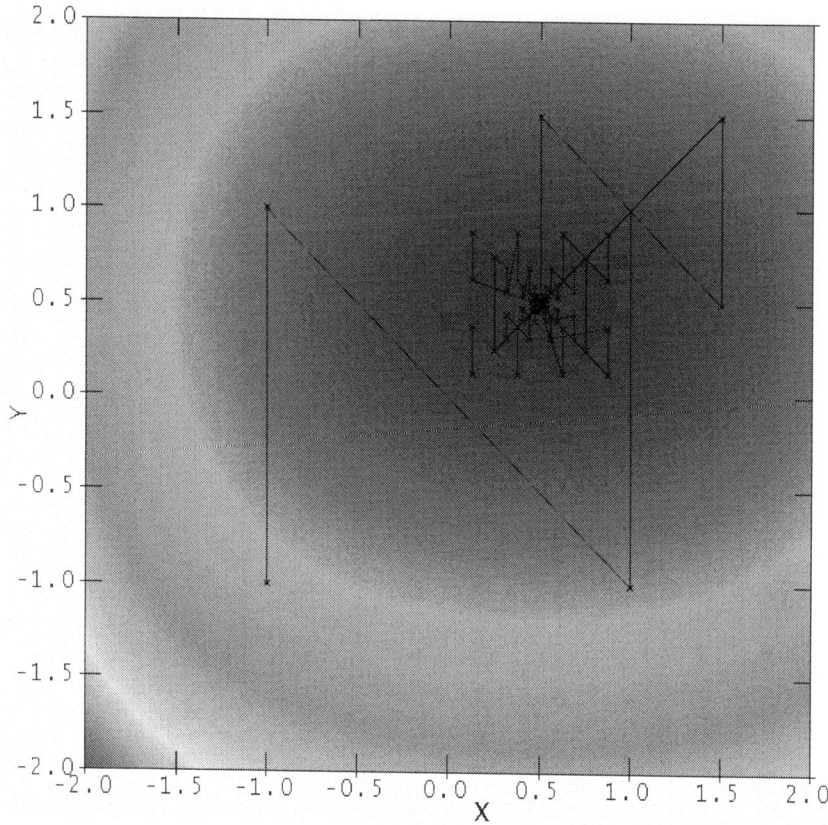

The last example of this technique we will consider is the first three-variable problem in Chapter 7: $f_1=15*x_1-cos(5/9*(x_2-1/5)*\pi*x_3)-1/2$, $f_2=25*x_1^2-81*(x_2-1/10)^2-sin(5*\pi*x_3/9)+1$, $f_3=exp(-5*x_1*(x_2-1/5))-100*\pi*x_3/9+10*\pi/3-1$. This requires expanding the technique to an arbitrary number of dimensions. The code (example3.c) must allocate and free memory every time it enters and leaves the function; otherwise, the values won't be preserved and the arrays will be orphaned in the heap.

Extension of this algorithm to an arbitrary number of dimensions requires a variable number of nested loops. While we could insert separate code for one loop, another for two, yet another for three, and so forth, but this would be tedious and unimaginative. What we need is a loop of loops! This is part of the test I give all prospective programmers. A professional should be able to implement it efficiently.

The most compact code performing an arbitrary loop of loops is the following:

```
do{
    for(i=0;i<n;i++)
        {
        if(k[i]<l[i]-1)
            {
            k[i]++;
            break;
            }
        k[i]=0;
        }
    }while(i<n);
```

Typical output is as follows:
```
loops 2 3 4
0 0 0 0
1 1 0 0
2 0 1 0
etc...
21 1 1 3
22 0 2 3
23 1 2 3
2*3*4=24
```

The loop of loops code, along with a batch file to compile and test it, can be found in the online archive in folder examples\bisection. We insert the loop of loops into the two-variable bisection code (example1.c and example2.c) to get the n-variable bisection code (example3.c). Then use this to solve the first three-variable problem from Chapter 7. Note the search pattern (1/4, 3/4), then split in half, and repeat:

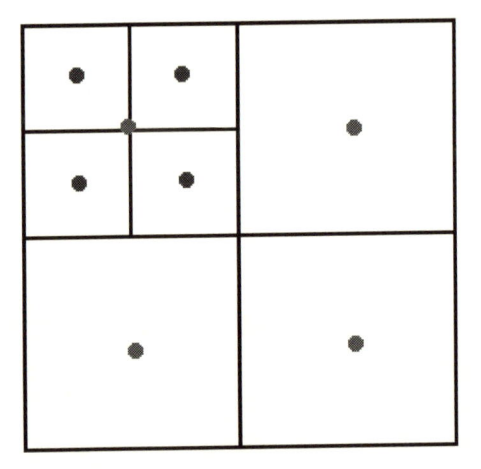

The 3D solution path for this problem is shown in the following figure:

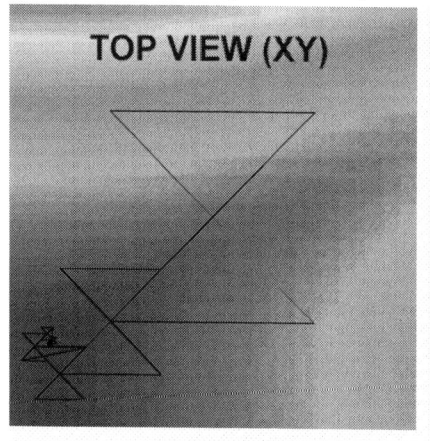

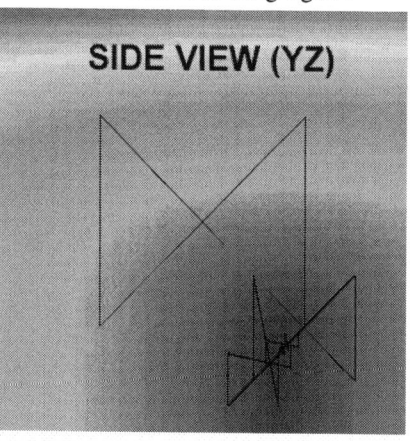

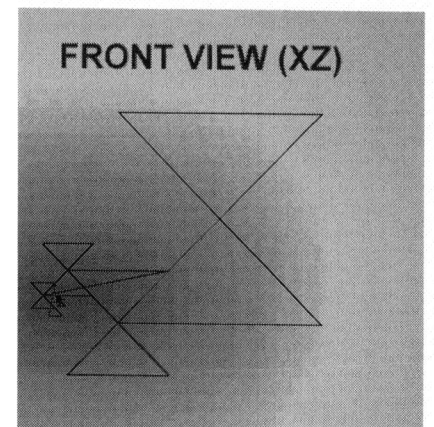

TOP VIEW (XY)

SIDE VIEW (YZ)

FRONT VIEW (XZ)

X=0.1
Y=0.2
Z=0.3

Chapter 10: Evolutionary Method

Evolutionary methods are based on the notion that accumulated small, random changes may eventually lead to something useful. The simplest possible implementation is to randomly hop around the domain, increasingly biasing the location with any fortuitous results. The following figure shows a one-dimensional example of this technique:

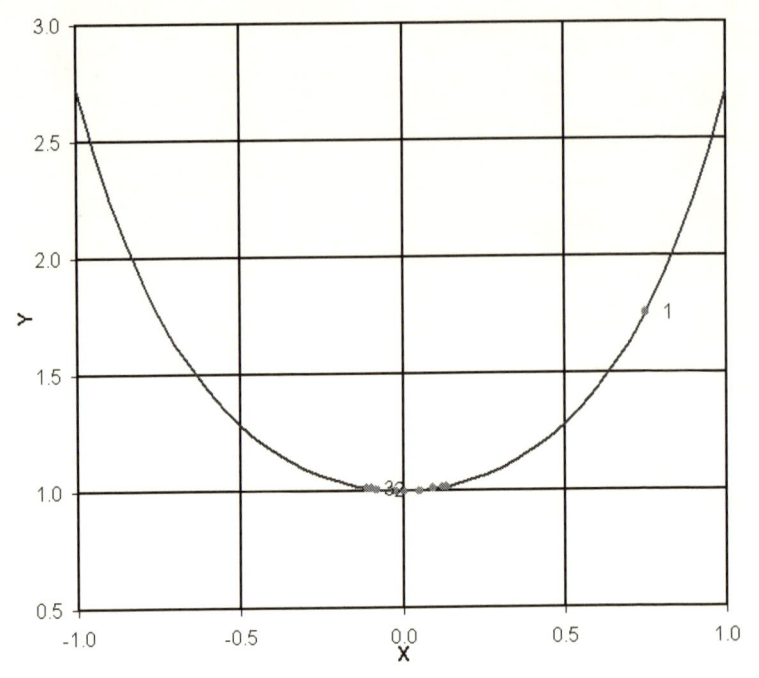

The method appears to work for this simple example. Of course, a rock could find this minimum.

The "bump" example from Chapter 3 is a little more challenging. It's shown in this next figure. The files can be found in the online archive in folder examples\evolutionary. This method isn't particularly efficient, but it's better than a completely random search.

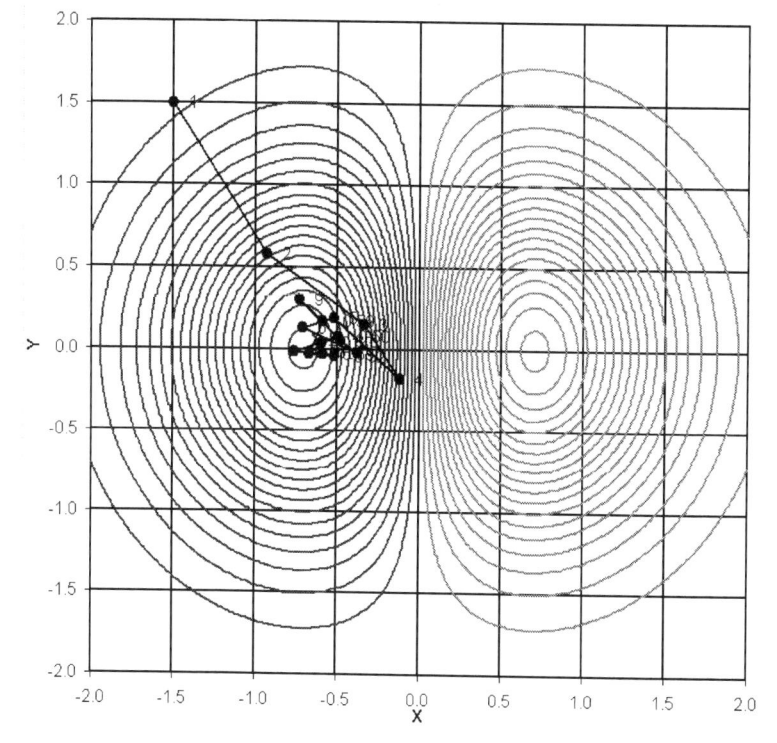

The third problem from Chapter 9 is an even greater challenge, as illustrated in this next figure:

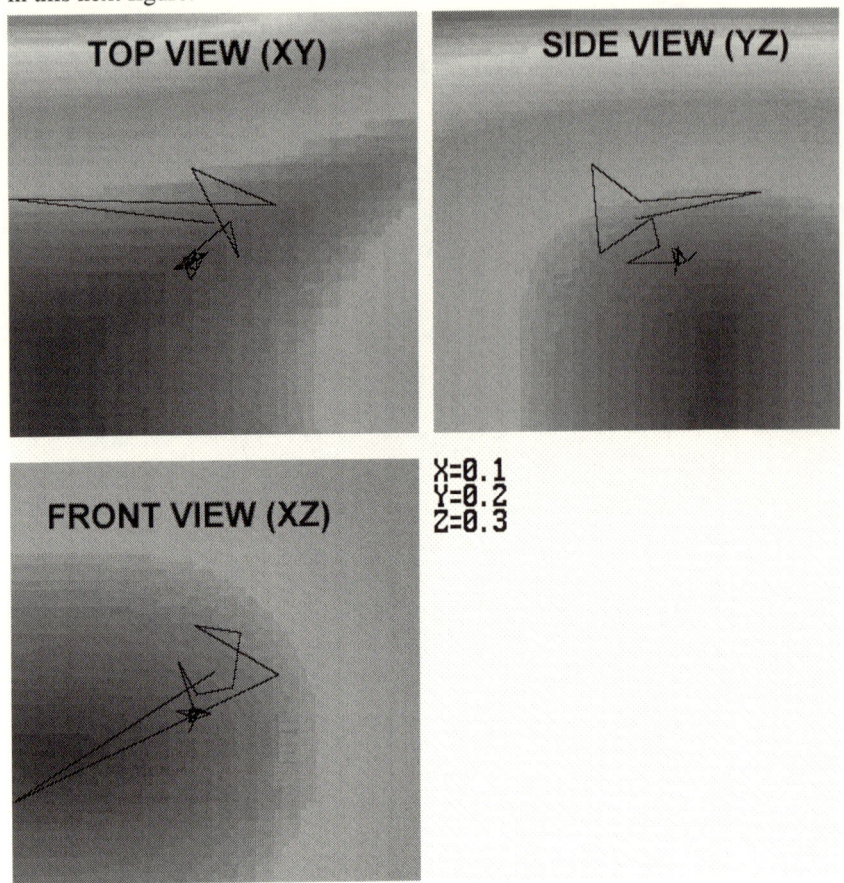

TOP VIEW (XY)

SIDE VIEW (YZ)

FRONT VIEW (XZ)

X=0.1
Y=0.2
Z=0.3

The method runs out of steam long before finding the blue core of the residual volume.

There are countless modifications, which might be applied to coax this method along. The simplest is to retard the influence of the previous solutions. This is illustrated in the next figure:

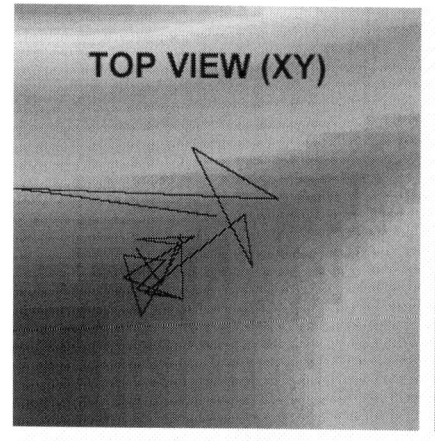

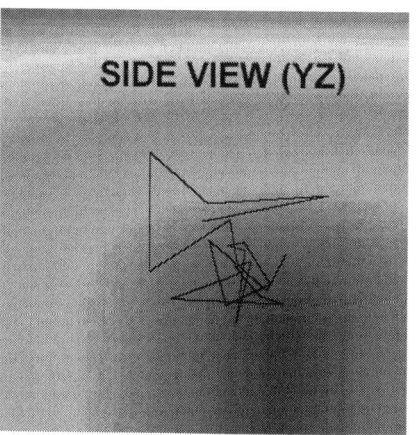

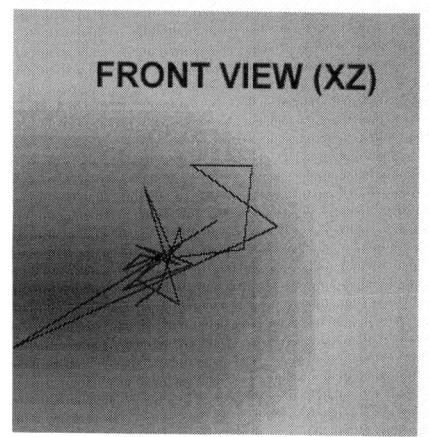

TOP VIEW (XY)

SIDE VIEW (YZ)

FRONT VIEW (XZ) X=0.1
Y=0.2
Z=0.3

Retardation helps a little in this particular case, but we want methods that work for most cases, not ones that must be endlessly tweaked for every new application. We will consider one more tweak before moving on to the next example. In this modification, we save the *n* best locations, where *n* is the number of independent variables or the dimension of the solution space and bias the guesses with this retained information about the residual. We allocate space for *n+1* residuals (*Fj*) an locations (*Xj*). We start with *n* random guesses and sort these on the residual. Each update goes in locations *Fj[n]* and *Xj[n]* and the list get resorted. The following code (example3a.c) implements this technique:

```
void EvolutionarySearch(double*Xm,double*Xx,int n,double
    func(double*),int m,double*X,double*F)
    {
```

```
int i,j;
double*Fj,*Xj;
Fj=calloc((n+1)*(n+2),sizeof(double));
Xj=Fj+n+1;
for(j=0;j<n;j++) /* n random initial guesses */
   {
   for(i=0;i<n;i++)
      Xj[n*j+i]=Xm[i]+rnd()*(Xx[i]-Xm[i]);
   Fj[j]=func(Xj+n*j);
   }
bubble_sort(Fj,Xj,n,n);
*F=Fj[0]; /* save best so far */
memcpy(X,Xj,n*sizeof(double));
for(;m>0;m--)
   {
   for(i=0;i<n;i++)
      {
      Xj[n*n+i]=n*(Xm[i]+rnd()*(Xx[i]-Xm[i])); /* random guess weighted with */
         for(j=0;j<n;j++) /* last n values with least Fjs */
            Xj[n*n+i]+=(n-j)*Xj[n*j+i];
         Xj[n*n+i]/=(n*(n+3))/2;
      }
   Fj[n]=func(Xj+n*n);
   bubble_sort(Fj,Xj,n,n+1);
   if(Fj[0]<*F)
      {
      *F=Fj[0]; /* save best so far */
      memcpy(X,Xj,n*sizeof(double));
      }
   }
free(Fj);
}
```

A bubble sort is good enough for our purposes here, as these arrays don't lend themselves to using the library function qsort(). The results are disappointing, as illustrated in the next figure. This modification utilizes more previous information and was allowed to run for five times as many iterations, yet failed to approach the solution. We have moved from one to two to three dimensions and already the Evolutionary Method seems to be petering out. This observation will be reinforced by subsequent examples comparing the gradient and evolutionary options of the Excel® Solver Add-In. Even after many, many more trial solutions, the gradient method stomps the evolutionary. Intelligence consistently beats brute force.

TOP VIEW (XY)

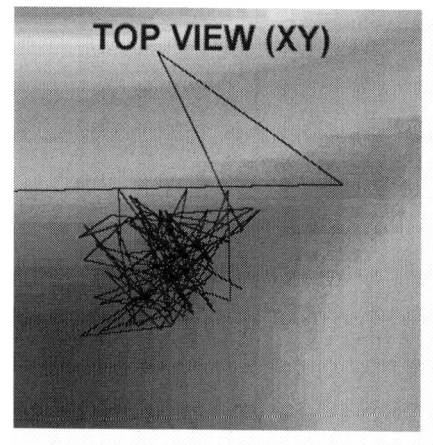

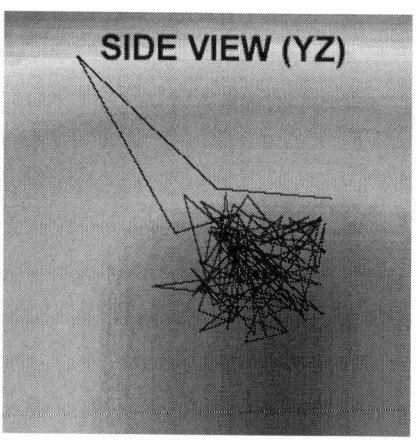

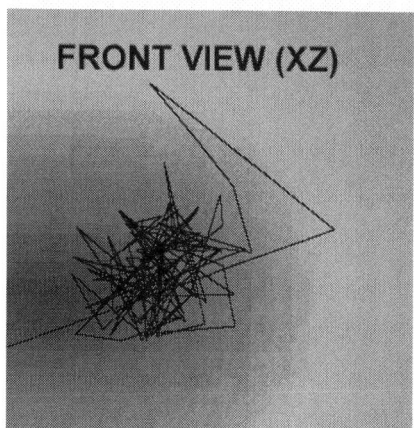

SIDE VIEW (YZ)

FRONT VIEW (XZ)

X=0.1
Y=0.2
Z=0.3

The Evolutionary Method quickly gets stuck in a region away from the solution where a steepest descent along the gradient would quickly lead to the optimal point. Of course, there are countless ways of restarting the process and trying over-and-over again, but you could apply those same "tricks" to one of the more intelligent methods (like I already did to Broyden's) and be ahead.

Chapter 11: Nonlinear Regression

In Chapter 5 we introduced nonlinear least squares, which is the basic strategy for nonlinear regression. In this chapter we will consider more complex problems. Linear regression is a non-iterative process in which matrices are built and solved, yielding a single result. Not that such are completely without complication, but iterative methods are much more complex and fraught with difficulties. We rarely approach a regression with no anticipation of the final form and so it is with the following examples.

Model Parameter Tuning

The first case we will consider is a mass transfer phenomenon that was killing fish in a similar way that the "bends" (i.e., decompression sickness) impacts divers. Some investigators thought this impossible, as fish move freely through the deep, but the biological evidence was overwhelming and persistent. The unusual situation impacting the fish was due to supersaturation of the water with nitrogen, immediately downstream of a dam.

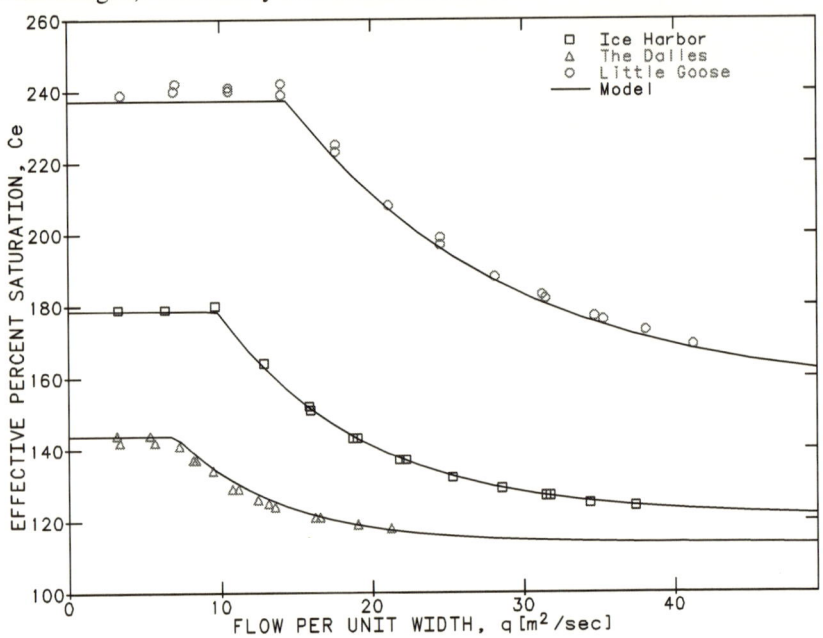

After collecting field data from four different sites (Ice Harbor, The Dalles, Little Goose, and Jennings-Randolph), a modeling effort was undertaken to better understand the mechanisms involved.[12] Supersaturation is a mass transfer

---

[12] Benton, D. J., "Modeling Nitrogen Supersaturation at Jennings-Randolph," U. S. Army Corps of Engineers Report. 1998.

process by which nitrogen from air bubbles enters the water and later tissues of the fish. Mass transfer processes are necessarily time-dependent and lead to the following solution:

$$C_d = C_s - (C_s - C_u)e^{-Kt} \tag{11.1}$$

$$C_e = C_s \left(1 + \frac{d_e \gamma}{P_a}\right) \tag{11.2}$$

$$d_e = h_2 + (h_1 - h_2)e^{\left(1 - \frac{\beta h_b}{L_s}\right)} \quad \text{for} \quad \frac{\beta h_b}{L_s} > 1$$

$$d_e = h_1 \quad \text{for} \quad \frac{\beta h_b}{L_s} \leq 1 \tag{11.3}$$

The heights, depths, and lengths are the physical dimensions of the dam and river The rate of change of the concentration is given by:

$$\frac{dC}{dt} = K_1 a_b (C_e - C) + K_2 a_s (C_s - C) \tag{11.4}$$

Here, K_1 and K_2 are mass transfer coefficients. The bubble mass transfer parameters are shown in this next figure:

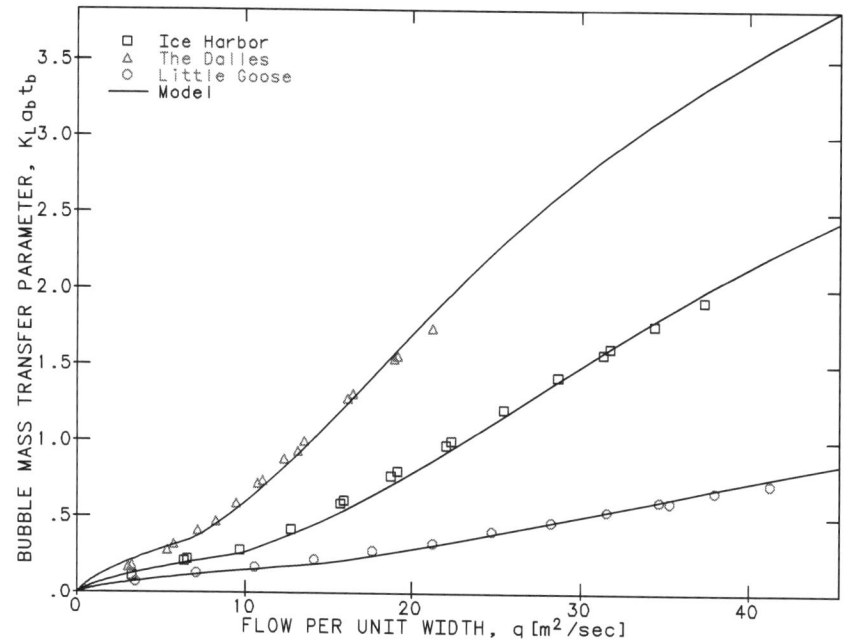

The solution to this differential equation is:

$$C_d = C_e - (C_e - C_u)\Omega$$
$$\Omega = e^{-(K_1 a_b t_b + K_2 a_s t_s)} + \Phi \left[1 - e^{-(K_1 a_b t_b + K_2 a_s t_s)}\right] \quad (11.5)$$
$$\Phi = \frac{(K_1 + K_2) a_s t_s}{K_1 a_b t_b + K_2 a_s t_s} \left(\frac{C_e - C_s}{C_e - C_u}\right)$$

The concentrations, C_d, C_e, and C_u, are unknown and will be adjusted to best fit the field data. As before, we have some idea what these must be, but the exact values are uncertain. This gives us bounds within which to search for an optimum solution, that is, one that best agrees with the available data. We will use Broyden's Method to find these three coefficients. All of the associated files can be found in the folder examples\nitrogen. The normalized optimum solution is shown in this next figure:

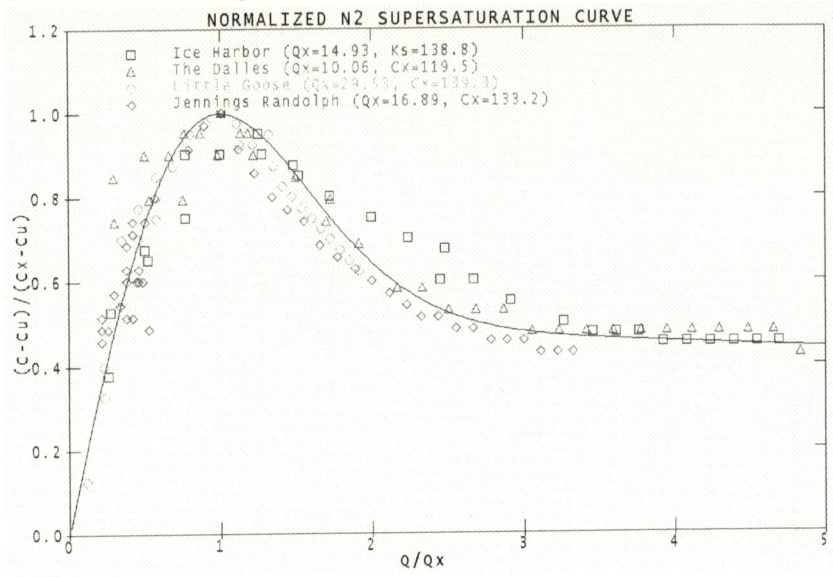

The residual function is simple and solved as in the previous examples using Broyden's Method:

```
double fun(double a,double b,double c,double x)
    {
    double C11,C12,C21,C22,D;
    C11=exp(-a)*erf(b);
    C12=erf(c);
    C21=exp(-a)*(a*erf(b)*sqrt(M_PI)-2*b*exp(-b*b));
    C22=-2*c*exp(-c*c);
    D=C11*C22-C21*C12;
    if(fabs(D)<FLT_MIN)
        Abort(__LINE__,"can't make match point");
```

```
A=C22/D;
B=-C21/D;
return(A*exp(-a*x)*erf(b*x)+B*erf(c*x));
}
```

The particular (non-normalized) optimum solution is shown in this figure:

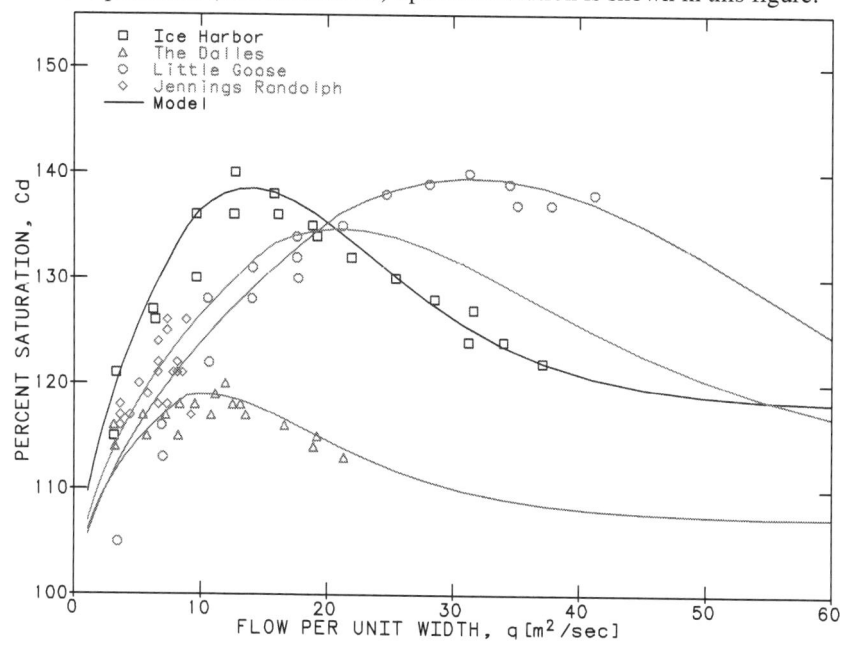

[Figure: Percent Saturation, Cd vs. Flow per unit width, q [m²/sec], with data points for Ice Harbor, The Dalles, Little Goose, Jennings Randolph, and Model curves]

In case you're wondering... changes were made and the fish are much happier. My report convinced the right people that this was indeed the problem. The preceding equations were used to optimize the reconstruction efforts.

Combined Cycle Power Plant Performance Data

We will now consider a greater challenge: a pair of nonlinear regression problems with 10 unknowns and 8562 residuals. We seek two approximations (power and heat input) based on one year of data from a typical modern combined cycle power plant. The data, code, and associated files can be found in the online archive in folder examples\power plant data.

In this case we have very definite expectations regarding the form of the regressions. There is a great body of work on power plant performance. Two foundation references are ASME PTC-46 (Overall Plant Performance) and PTC-22 (Gas Turbines). Equipment manufacturers and plant designers all over the world follow these guidelines. Original equipment performance is most often provided as laid out in these two documents. Power and heat input are expected to have the following form:

$$\frac{P_{REF}}{P} = \prod_{i=1}^{14} \alpha_i \qquad (11.6)$$

$$\frac{Q_{REF}}{Q} = \prod_{i=1}^{14} \beta_i \qquad (11.7)$$

The power, P, and heat input, Q, are normalized by some reference value, typically the guarantee at the base reference conditions. The most common base reference conditions are 1 atmosphere (corrected for local elevation), 59°F (15°C), and 60% relative humidity. The multiplicative corrections, α_1 through α_{14} and β_1 through β_{14}, account for deviations from the base reference conditions. Each index corresponds to a specific parameter, as listed in the following table.

In this case we will only use the ones in bold: temperature, pressure, humidity, fuel composition, speed, equivalent hours of operation (EOH), and inlet guide vane position (indicative of part-load operation). The last two are not allowed in a contractual acceptance test, when demonstration of full-load operation in "new and clean" condition is the objective. Two can be quickly simplified: pressure and speed, as capacity (power) and consumption (heat input) are directly proportional to these. All of the corrections must be unity (i.e., equal to one) at the base reference conditions.

1	**inlet air temperature**
2	**barometric pressure**
3	**inlet humidity**
4	**fuel composition**
5	injection fluid flow
6	injection fluid enthalpy
7	injection fluid composition
8	exhaust pressure loss
9	**shaft speed**
10	turbine extraction
11	fuel temperature
12	inlet pressure loss
13	**hours of operation**
14	**inlet guide vane position**

The impact of temperature on gas turbine (and thus combined power plant) performance varies with design, but generally follows the same trend. These (and every other heat engine) perform better at low temperatures than high. As these corrections are the inverse, this means that α_1 is greater than 1 at low temperatures and less than 1 at high temperatures, as illustrated below:

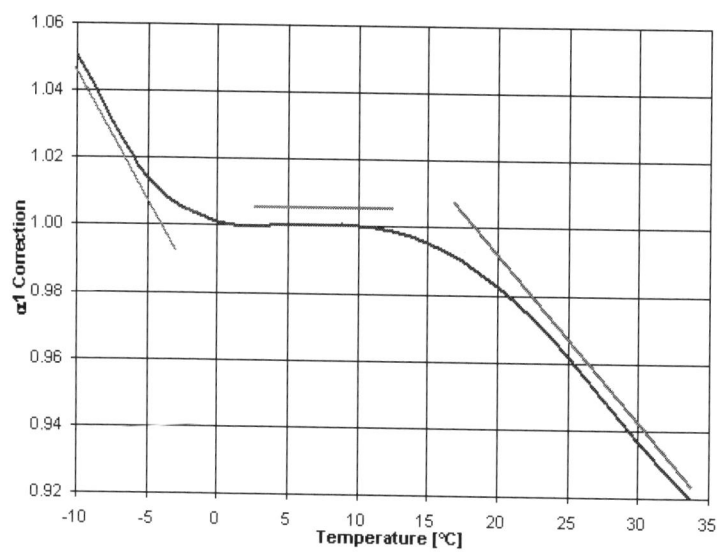

There is often (but not always) a flat spot in the middle of this curve. It is very important that the asymptotic behavior at both ends of the curve exhibit a negative slope. This means, if we are to approximate the shape of this curve as a polynomial in temperature, it must have an odd order of at least three. We rarely exceed order three, as the curve may flip around—something to avoid.

Relative humidity is a familiar term, but most people don't appreciate the sizeable impact water vapor content has on air. We all know that hot and humid is more uncomfortable than hot and dry. At 70°F (21°C) and 100% relative humidity, there is more energy in the water vapor than the dry air. At 120°F (49°C) and only 40% relative humidity, this is also true. The water vapor bearing capacity of air changes vastly over the annual range of ambient temperatures in most locations.

The impact of relative humidity on gas turbine (and thus combined cycle) performance is highly nonlinear and strongly dependent on temperature; however, the impact of *absolute* humidity is linear. Relative humidity is the fraction of water vapor compared to the maximum (i.e., bearing capacity); whereas, absolute humidity is simply the mass fraction, irrespective of the bearing capacity. This is why it's always preferable to base such regressions on absolute, rather than relative, humidity.

The impact of fuel composition (the chemistry of natural gas varies with the source) is most often adequately approximated as being linear in heating value and carbon to hydrogen ratio. The impact of ageing (equivalent hours of operation) is most pronounced early in the life of an engine and flattens-out over time. This particular engine has been running for several years, so we will assume it to be on the flat (i.e., linear) region.

Heat rate is the ratio of the heat input to the power and is inversely proportional to the thermal efficiency. Heat rate is a common measure of performance for power plants. Heat rate has considerably greater scatter than either power or heat input, which is why these two are calculated or measured before taking the ratio and why we will base our second regression on heat input rather than heat rate.

The corrections are as follows:

$$\alpha_1 = C_0 + C_1 T + C_2 T^2 + C_3 T^3 \quad (11.8)$$

$$\alpha_3 = C_0 + C_1 humidity \quad (11.9)$$

$$\alpha_4 = C_0 + C_1 LHV + C_2 CH_{RATIO} \quad (11.10)$$

$$\alpha_{13} = C_0 + C_1 EOH \quad (11.11)$$

$$\alpha_{14} = C_0 + C_1 IGV + C_2 IGV^2 \quad (11.12)$$

The β corrections for heat input have the same form. Each of these equations can be simplified by using the difference between actual and base reference quantities, which makes C_0 equal to 1, for instance, in the case of temperature:

$$\Delta_T = \frac{T - T_{BASE}}{T_{BASE}} \quad (11.13)$$

We also know from experience that inverting the 13th and 14th corrections results in less scatter, so that the final complete form of the regression becomes:

$$P = C_0 \left(\frac{baro}{baro_{BASE}}\right)\left(\frac{speed}{speed_{BASE}}\right) \times \frac{(1 + C_4 \Delta_{LHV} + C_5 \Delta_{CH_{RATIO}})(1 + C_6 \Delta_T + C_7 \Delta_T^2 + C_8 \Delta_T^3)(1 + C_9 \Delta_{HUM})}{(1 + C_1 \Delta_{EOH})(1 + C_2 \Delta_{IGV} + C_3 \Delta_{IGV}^2)} \quad (11.14)$$

This gives us 10 parameters to adjust in order to minimize the residual (i.e., sum of the squares of the discrepancies) with respect to the reported data. The regression for heat input has the same form. This problem is nonlinear in that the coefficients multiply each other. We can separate the two groups on the bottom of Equation 11.14, thus:

$$P(1 + C_1 \Delta_{EOH})(1 + C_2 \Delta_{IGV} + C_3 \Delta_{IGV}^2) baro_{BASE} speed_{BASE} = baro \times speed \times \quad (11.15)$$
$$C_0 (1 + C_4 \Delta_{LHV} + C_5 \Delta_{CH_{RATIO}})(1 + C_6 \Delta_T + C_7 \Delta_T^2 + C_8 \Delta_T^3)(1 + C_9 \Delta_{HUM})$$

This doesn't eliminate the nonlinearity. In fact, expanding Equation 11.15 results in 30 terms. Solving the problem using linear regression would require finding 30 coefficients and then requiring 20 combinations thereof to have certain behavior such that that the whole would reduce back down to Equation 11.15. As this would also be a nonlinear problem, we solve the problem as is.

The Excel® Solver Add-In was first used to determine the 10 coefficients. The Solver takes several minutes to converge to a solution with a residual of $10^{5.005}$ for power and $10^{6.541}$ for heat input. Broyden's derivative-free method (see file plant_data.c) is much faster (0.3 second) and achieves the same residual. The Solver algorithm is a Generalized Gradient Reduction (GRG) method, using finite differences to calculate the various partial derivatives. Of course, any native executable will run a lot faster than Excel®.

After running for about 1 second (the native, not Excel® version, which takes several minutes), the Evolutionary Method achieved a residual of $10^{6.309}$. After running for over 9 seconds the residual was finally reduced to $10^{5.775}$. Results of the power regression are shown in the following figure and of the heat input regression are shown in the one after that. Either solver can be selected in Excel®. The multi-dimensional bisection method achieved a residual of $10^{6.309}$ after running for about 9 seconds. All three methods are included in the source code.

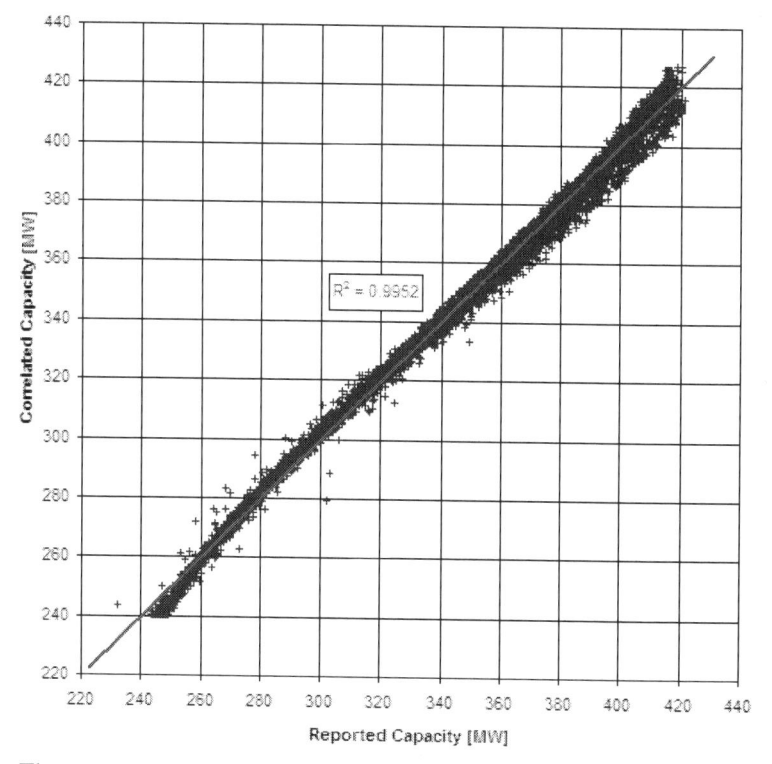

The program output is:

```
examples\power plant data>plant_data
reading data: plant_data.csv
  8563 lines read
  8562 data points found
Power Regression
  Uncorrelated Data
    log10(residual)=9.017
  Excel GRG Solver
    log10(residual)=5.005
  Bisection Search
    9216 calls 8.507 seconds
    log10(residual)=5.555
  Broyden's Method
    100 calls 0.333 seconds
    log10(residual)=5.006
  Evolutionary Method
    1009 calls 0.942 seconds
    log10(residual)=6.309
  Evolutionary Method
    10009 calls 9.429 seconds
    log10(residual)=5.775
Heat Input Regression
  Uncorrelated Data
    log10(residual)=5.775
  Excel GRG Solver
    log10(residual)=6.541
  Bisection Search
    9216 calls 8.623 seconds
    log10(residual)=7.222
  Broyden's Method
    79 calls 0.225 seconds
    log10(residual)=6.542
  Evolutionary Method
    1009 calls 0.939 seconds
    log10(residual)=7.816
  Evolutionary Method
    10009 calls 9.304 seconds
    log10(residual)=7.453
```

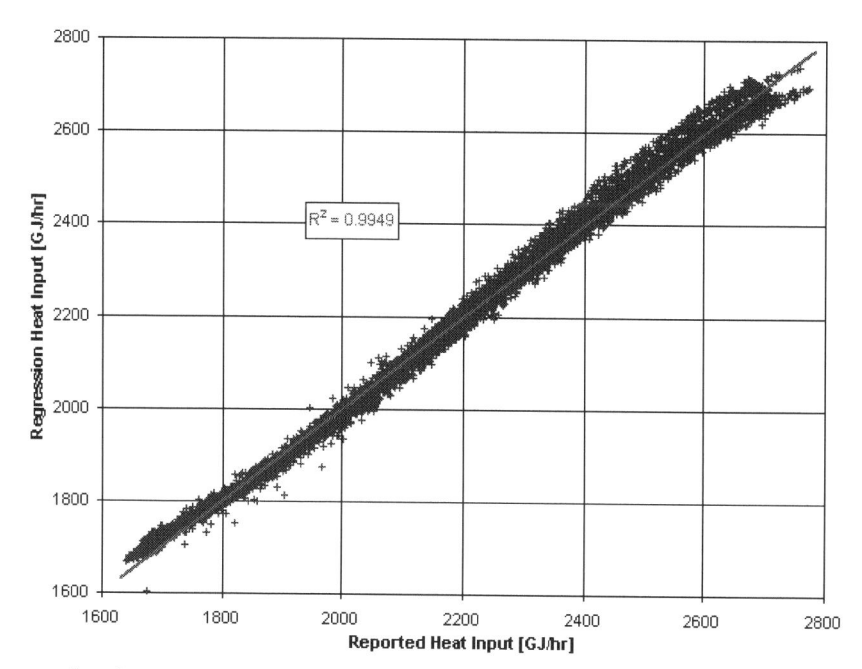

An important principle is illustrated by these examples: simplistic techniques based on wishful thinking or dependant on brute force are rarely preferable to a thoughtful and reasoned approach. One application where this wisdom has been embraced is computer chess. The most successful programs don't try to consider every possible move; rather, they use strategy. You will find that it take quite a while for the Evolutionary Method to run in Excel®. Try it and see!

Pressure-Temperature-Density Data

The next example we will consider is the pressure, temperature, and density behavior of real fluids. Every pure substance that is chemically stable over a sufficient range of pressure and temperatures exhibits both a critical and triple point. The critical point is where the liquid and vapor are physically indistinguishable. The triple point is where the solid, liquid, and vapor coexist in equilibrium. The critical point of water occurs at 22 MPa (3200 psia) and 374°C (705°F). The triple point occurs at 611 Pa (0.89 psia) and 0°C (32°F).

GENERALIZED COMPRESSIBILITY FACTORS (Zc=0.27)

compressibility factor, z

reduced pressure, P_r

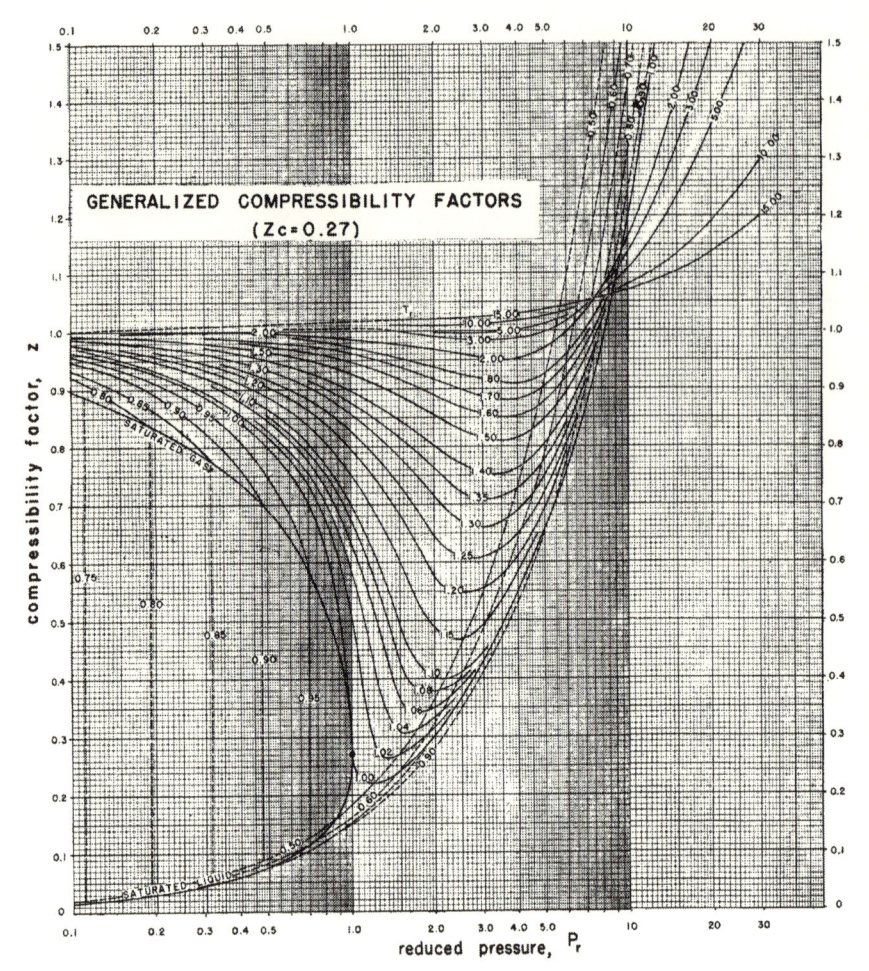

    The ideal gas law is: $PV=RT$, where R is the ideal gas constant. For non-ideal states we define the compressibility factor, $Z=PV/RT$. Nelson and Obert introduced the generalized compressibility chart in 1954[13]. The preceding figure shows compressibility, Z, vs. reduced pressure, $Pr=P/Pcritical$, for various values of reduced temperature, $Tr=T/Tcritical$. I have digitized this figure and put the data into an Excel® spreadsheet in folder examples\compressibility. Fitting a single function to this entire family of curves might seem like an insurmountable task, but that's precisely what we're going to do.

---

[13] Nelson, L. C. and Obert, E. F., "Generalized pvT Properties of Gases," Transactions of the ASME, p. 1057, October, 1954.

The first equation of state to account for real fluid behavior was introduced by van der Waals in 1873: $P=RT/(V-b)-a/V^2$. V is the specific volume, equal to 1/density. The two constants, a and b, are selected to match the "bulge" in the preceding figure at $Pr=1$, $Z=0.27$. In order to form the "bulge" the following must be true:

$$\frac{\partial P}{\partial V} = \frac{\partial^2 P}{\partial V^2} = 0 \bigg|_{P=Pc, V=Vc} \tag{11.16}$$

While the van der Waals equation has somewhat the correct shape, it does not agree well with data, especially for liquids. We seek an improvement upon this first attempt at capturing real fluid behavior. The simplest improved form is:

$$P = \frac{RT}{V-b} - \frac{a}{\left(V^2 + cV + d^2\right)} \tag{11.17}$$

We will allow a, b, c, and d to be temperature-dependent. Examining the data reveals that a, c, and d should all get smaller with increasing temperature, while b should get smaller. Therefore, we propose the following relationships:

$$a, c, d \propto Tr^n, n > 0$$
$$b \propto Tr^n, n < 0 \tag{11.18}$$

This gives us 4 constants plus 4 exponents to adjust so as to minimize the residual. Experience also reveals that trying to match pressures at the same specific volume (1/density) yields spurious and problematic results. Instead, we want to match specific volume at the same pressure. This means that all of the calculations are implicit, as the X-axis of the figure is pressure, not specific volume. It takes some coding and it runs slowly, but Excel® can handle it and arrives at the following results. The blue specks are the digitized points and the red curves are the regression.

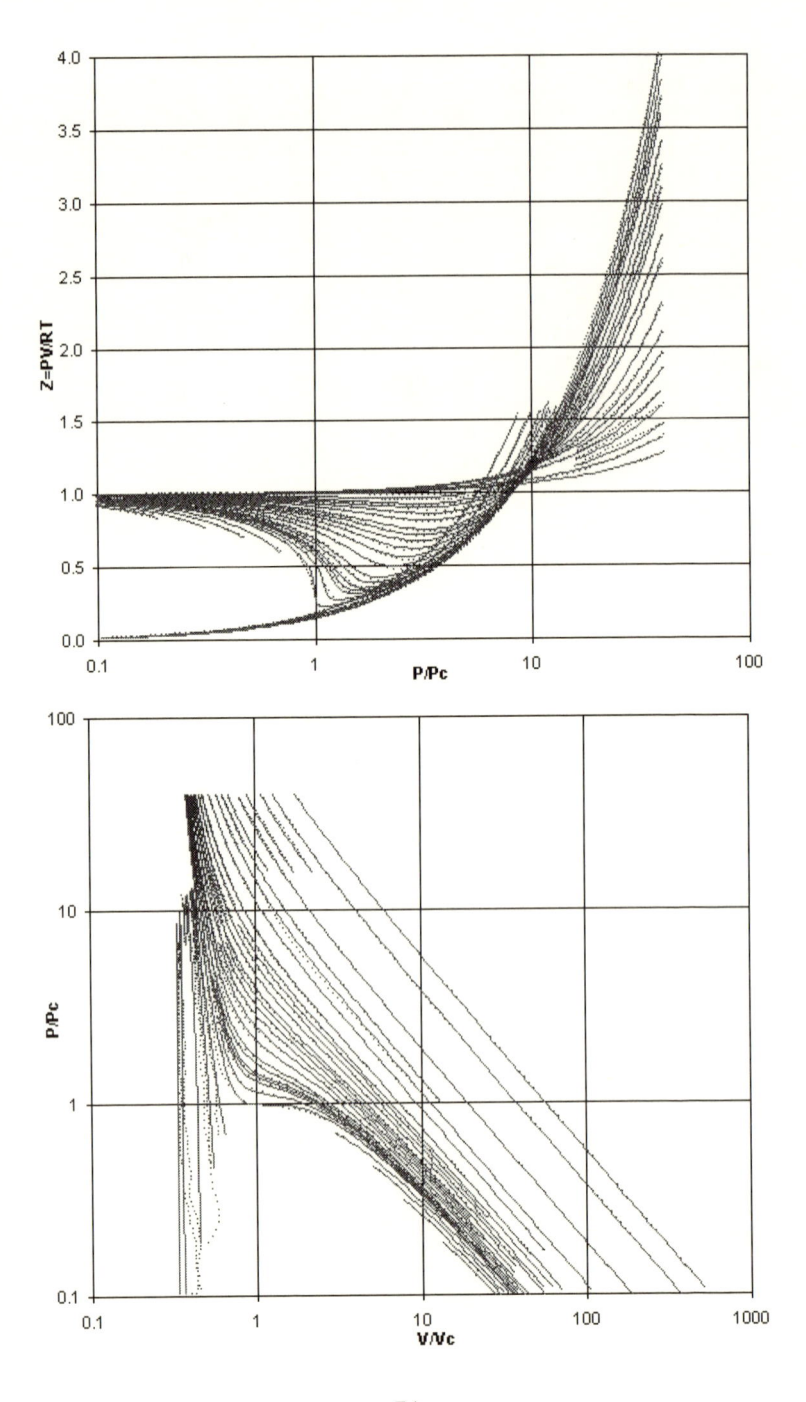

74

The second figures shows the same data on a different axis, this time reduced pressure vs. reduced volume. On the second figure, the critical point is at location (1,1) and also appears as a "hump" of sorts. Considering the X-axis in both figures is logarithmic, as is the Y-axis on the second figure, the agreement for such a simple equation is remarkable.

We apply the same techniques as for the power plant data (see source code in file compressibility.c). The program outputs:

```
examples\compressibility>compressibility
reading data: compressibility.csv
  3420 lines read
  3419 data points found
Uncorrelated Data
  residual=830114.310
Excel GRG Solver
  residual=1.019
Bisection Search
  2304 calls 7.143 seconds
  residual=7.643
Broyden's Method
  100 calls 0.386 seconds
  residual=1.060
Evolutionary Method
  1007 calls 3.056 seconds
  residual=4.223
Evolutionary Method
  10007 calls 30.391 seconds
  residual=1.304
```

Again, Broyden's Method is much faster than the Excel® GRG Method and yields equivalent results. The Bisection Method is slow and yields somewhat disappointing results. The Evolutionary Method is even slower and yields disappointing results.

Viscosity

Like thermodynamic properties, transport properties are essential to solving many scientific and engineering problems. The viscosity of liquids and vapors varies considerably over the range of operating temperatures. Perhaps the most studied fluid is water. As with preceding example, a high-order linear regression is unlikely to have the character we desire or the correct asymptotic behavior; thus, we turn again to nonlinear regression and an informed choice of form.

The following graph shows the dynamic viscosity of water in kg/s/m x10^{-6}, which is milli-centi-poise. Other than being convenient numbers ranging up to 1000, I have no idea why anyone would use such a combination of metric units. In any event, divide by 413.3788731 to get the far more useful units of lbm/ft/hr. You can find all of the related files in folder examples\viscosity.

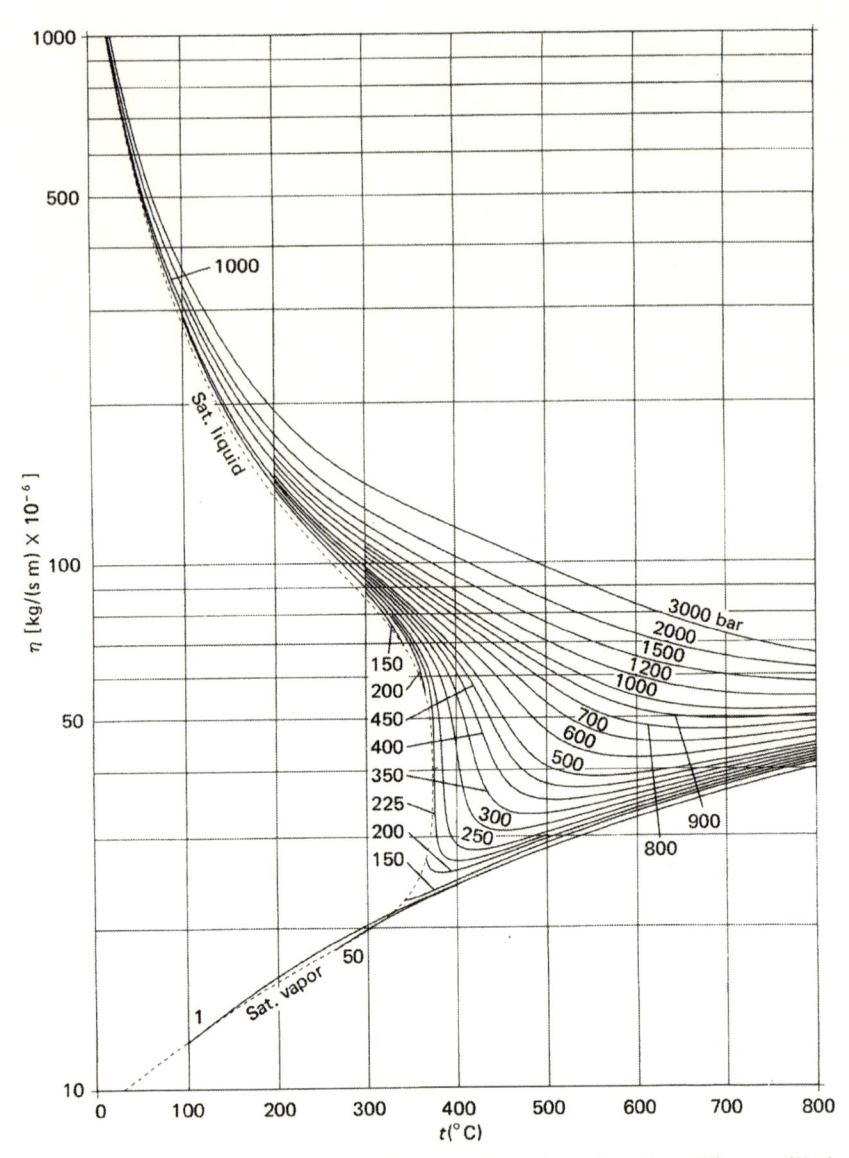

From working with various fluids, we expect that viscosity will more likely be related to density than either temperature or pressure, so we transform the X-axis into density [gm/cm²] and re-plot to obtain the following, much more tractable regression problem:

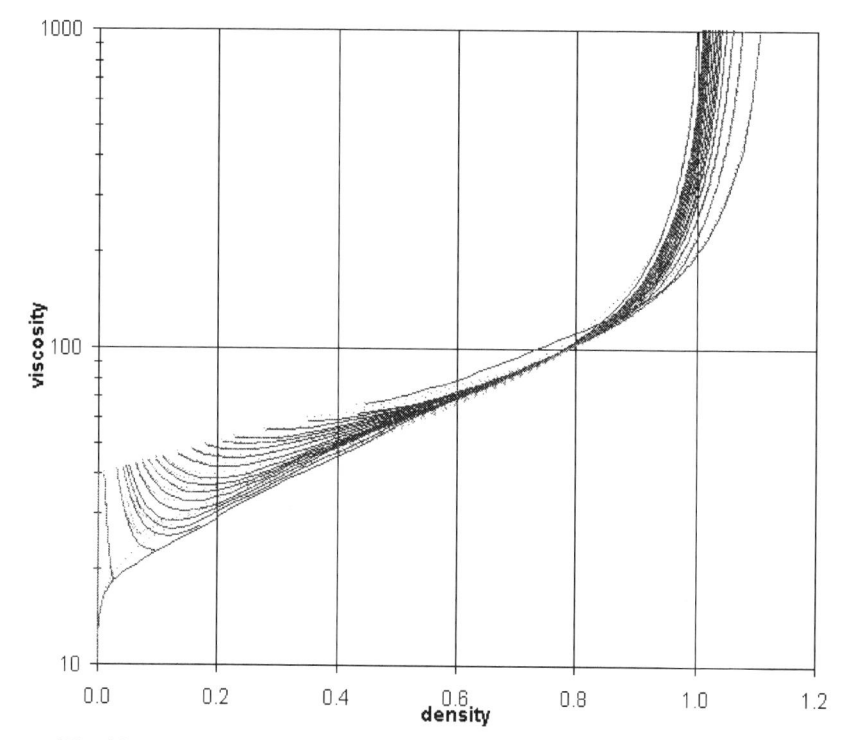

[Graph: viscosity vs density]

The blue curves are data from the preceding graph and the orange specks are the regression. Again, this is surprisingly good. The basic form is:

$$\mu = \frac{\alpha}{\delta - \rho}$$
$$\alpha = f(P,T,\rho) \tag{11.19}$$
$$\delta = f(P)$$

After trying several formulas, we arrive at the following expressions for α and δ:

$$\alpha = C_1 + C_2 P + C_3 T + C_4 \rho + C_5 P^2 \ldots$$
$$+ C_6 PT + C_7 P\rho + C_8 T^2 + C_9 T\rho + C_{10} \rho^2 \tag{11.20}$$
$$\delta = C_{11} + C_{12} P + C_{13} P^2$$

We adjust the 13 coefficients to minimize the residual. In this case, we want to minimize the relative discrepancies and so we calculate the residuals:

$$r_i = \frac{fit - data}{data} \tag{11.21}$$

The same behavior would result if taking the difference in the logs. We don't use the logs because Excel® gets upset when taking the log of a negative value and such may arise during the solution process. It takes a few minutes, but Excel® settles in on the solution displayed in the previous graph. The program (viscosity.c) is very similar to the previous (compressibility.c). The output is:

```
examples\viscosity>viscosity
reading data: viscosity.csv
  2289 lines read
  2287 data points found
Uncorrelated Data
  residual=2287.000
Excel GRG Solver
  residual=2.702
Bisection Search
  114688 calls 16.609 seconds
  residual=575.417
Broyden's Method
  100 calls 0.080 seconds
  residual=265.087
Broyden's second chance
  100 calls 0.083 seconds
  residual=3.281
Evolutionary Method
  1012 calls 0.154 seconds
  residual=8.806
Evolutionary Method
  10012 calls 1.460 seconds
  residual=4.704
```

Broyden's Method doesn't work well at all for this case (residual 265 vs. 2.7 for Excel's GTG solver) in spite of being fed a reasonable facsimile of the answer because the solution domain is radically split up where $\mu = a/(\delta - \rho)$ blows up. This method will find a solution if such regions are eliminated, as with the second chance above.

Chapter 12: Hybrid Regression

Sometimes part of the regression can be done linearly and part cannot. In this case, we repeatedly perform the linear least squares part and adjust the rest to minimize some residual. The problem we will consider is based on Case 4 in the SIAM paper. The objective is to develop a simple approximation that will reasonably approximate the transient behavior of a reservoir, including the time lag from the upstream and downstream dams. The upstream and downstream flows come from actual operations at the dam. The calculated flows come from a reservoir routing model, which is too large and takes too long to run to be embedded in the decision management software; thus, a simplified solution is required.

The simplified solution approximates the calculated flow from a linear combination of 4 flows: the current upstream and downstream value plus two previous flows, one from upstream and one from downstream. The time lag for each is sought so as to best approximate the actual flow. The time lag is presumed to be at least 1 and no more than 24 hours. The time lags for the upstream and downstream are presumed to be different.

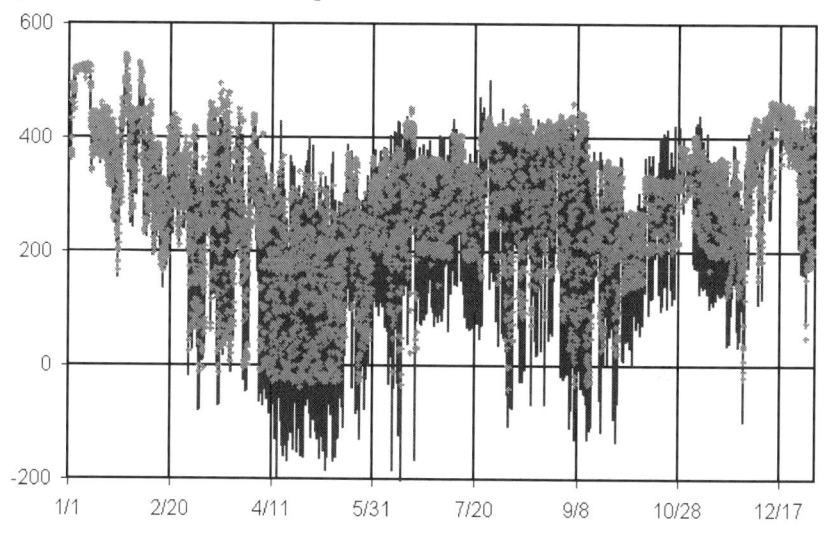

Integer lags are assumed in the Excel® spreadsheet (examples\ reservoir\ reservoir.xls) to facilitate solution, although this isn't strictly necessary. Excel's GRG solver doesn't work with this problem because it can't calculate the gradient of the residual by finite difference with respect to the time lags. Excel's Evolutionary solver does work, producing an answer of 2.33 (rounded to 2) and 6 hours for the upstream and downstream flows, respectively. The final residual is 15,407,749. I have provided a button to run through 24x24 cases sequentially, saving the best one, which turns out to be 3 and 7, with a residual of 12,682,871.

The Evolutionary solver actually takes longer than the brute force double loop, as it tries more combinations.

The end result is shown in the preceding figure. The blue lines are the flows from the routing model and the orange specks are the approximation. We will implement this in C (reservoir.c) without restricting the time lags to be integers. This would be possible in Excel® but impractical. The program output is:

```
examples\reservoir>reservoir
reading data: reservoir.csv
  8784 lines read
  8784 data points found
  Uncorrelated Data
  log(residual)=8.943
begin brute force (loop) solution
  1 1 18237780
  2 1 14481130
  3 1 12708036
  3 6 12695179
  3 7 12682871
  576 calls 0.218 seconds
  log(residual)=7.103
Excel GRG Solver
  log(residual)=7.103
Bisection Search
  12 calls 0.006 seconds
  lu=3.9, ld=21.1, log(residual)=7.173
Broyden's Method
  100 calls 0.088 seconds
  lu=4.9, ld=8.9, log(residual)=7.107
Evolutionary Method
  1001 calls 0.435 seconds
  lu=3.1, ld=7.4, log(residual)=7.099
Evolutionary Method
  10001 calls 4.392 seconds
  lu=3.2, ld=19.1, log(residual)=7.122
```

All of the methods work fairly well and run a *whole lot faster* than Excel®.

Appendix A: Steam Properties - A Practical Example

It's easy to provide trivial examples tailored to accentuate the strengths of one method over another, thus rigging the outcome. It is quite another thing to provide meaningful applications. We will now consider one such example: thermodynamic properties of steam. Properties are calculated many times in a spreadsheet or model. Each calculation involves a length function call with nested loops and transcendental functions, thus it can take a significant amount of time, even on modern computers.

No one has used Keenan, Keyes, Hill, and Moore's (KKHM) 1969 steam properties[14] for decades. They were never widely used, even when they first came out. They were a little more accurate than the 1967 properties[15], which had been endorsed by the ASME and still used by the General Electric Steam Turbine Division to this day (ASME67). There are three more sets of steam properties you should know about: the 1984, published by the NBS/NRC[16] (the National Bureau of Standards is now called the National Institute of Standards and Testing), the IAPWS-SF95[17], and the IAPWS-IF97.[18]

The KKHM properties are based on an elegant formulation and for that reason alone are of significant historical value. The NBS/NRC took this one step farther. It is a great shame that these two works have fallen into disuse. The IAPWS-SF95 is also an elegant formulation, though this work would not have been possible without the trailblazing authors of KKHM and NBS/NRC. The ASME67 and IAPWS-IF97 are both mathematical atrocities and should have been tossed in the recycle bin long ago.

Why are the steam properties so computationally intensive? Because we most often know temperature and pressure, but these are not always independent variables and, therefore, can't adequately represent the behavior of a substance over a wide range of conditions. All chemically-stable exhibit saturation stated in which the liquid or solid and vapor are in equilibrium. These same substances also exhibit two very interesting points: 1) the critical point, at which the liquid

---

[14] Keenan, J. H., Keyes, F. G., Hill, P. G., and Moore, J. G., *Steam Tables*, John Wiley & Sons, Inc., 1969.
[15] Meyer, C. A., McClintock, R. B., Silvestri, G. J., and Spencer, R. C., Jr., *Thermodynamic and Transport Properties of Steam*, American Society of Mechanical Engineers, 1967.
[16] Haar, L., Gallagher, J. S., and Kell, G. S., *Steam Tables*, NBS/NRC printed by Hemisphere, distributed by McGraw-Hill, 1984.
[17] Friend, D. G. and Dooley, R. B., *Revised Formulation for the Thermodynamic Properties of Ordinary Water Substance for General and Scientific Use*, The International Association for the Properties of Water and Steam, 1995
[18] Research and Technology Committee on Water and Steam in Thermal Power Systems, *ASME Steam Properties for Industrial Use*, The American Society of Mechanical Engineers.

and vapor are indistinguishable, and 2) the triple point, at which the solid, liquid, and vapor coexist in equilibrium.

Temperature and density are always independent for such substances and the properties are continuous in these two variables. Therefore, the formulation must also be continuous in these two variables. This is why the ASME67 and IAPWS-IF97 are mathematical atrocities—they aren't continuous and are jumble of brute-force curve fitting. Keenan, Keyes, Hill, and Moore weren't the first to publish it, but they were the first to implement this important observation: all of the thermodynamic properties of a substance can be expressed in terms of the Helmholtz Free Energy (HFE) or its derivatives. Find the HFE ($\psi=u-Ts$) and you have the rest!

KKHM began by splitting the HFE into two parts: one at zero density (i.e., rarified gas) plus a second at finite density that vanishes as the density approaches zero. The first part is:

$$\psi_0 = \sum_{i=1}^{8} \frac{C_i}{\tau^{i-1}} + C_7 \ln(T) + C_8 \frac{\ln(T)}{\tau} \tag{A.1}$$

where τ is a dimensionless inverse of the temperature:

$$\tau = \frac{1000}{T} \tag{A.2}$$

where T is the temperature in °K. The whole expression for HFE, ψ, is:

$$\psi = \psi_0 + RT(\ln \rho + \rho Q) \tag{A.3}$$

where ρ is the density in gm/cm³, R is the ideal gas constant (0.46151 J/gm/°K), and Q is the *partition* function, which is:

$$Q = (\tau - \tau_c) \left\{ \sum_{j=1}^{7} (\tau - t_j)^{j-2} \left[\sum_{i=1}^{8} A_{i,j} (\rho - r_j)^{j-1} + e^{-E\rho} \sum_{i=9}^{10} A_{i,j} \rho^{i-9} \right] \right\} \tag{A.4}$$

where $t_j=\tau_c$ for j=1, otherwise t_j=2.5, and rj=0.634 for j=1, otherwise 1. Pressure is given by:

$$p = \rho RT \left(1 + \rho Q + \rho^2 \frac{\partial Q}{\partial \rho} \right) \tag{A.5}$$

Enthalpy is given by:

$$h = RT \left(\rho \tau \frac{\partial Q}{\partial \tau} + 1 + \rho Q + \rho^2 \frac{\partial Q}{\partial \rho} \right) + \frac{d(\psi_0 \tau)}{d\tau} \tag{A.6}$$

Entropy is given by:

$$s = -R\left(\ln\rho + \rho Q - \rho\tau\frac{\partial Q}{\partial \tau}\right) - \frac{d\psi_0}{dT} \quad \text{(A.7)}$$

Equations A.1 through A.7 require temperature and density as inputs, yet we most often know temperature and pressure or some other property (saturated liquid, saturated vapor, enthalpy, or entropy). This means that we must solve these lengthy equations implicitly for density over and over again. The precise value of density is rarely of concern; however, an imprecise value of density yields an inaccurate value of pressure, enthalpy, or entropy, which is why we must always obtain an accurate solution.

Solving for the density over the entire range of interest is not as easy as it may seem. The following figure is commonly used to illustrate the problems:

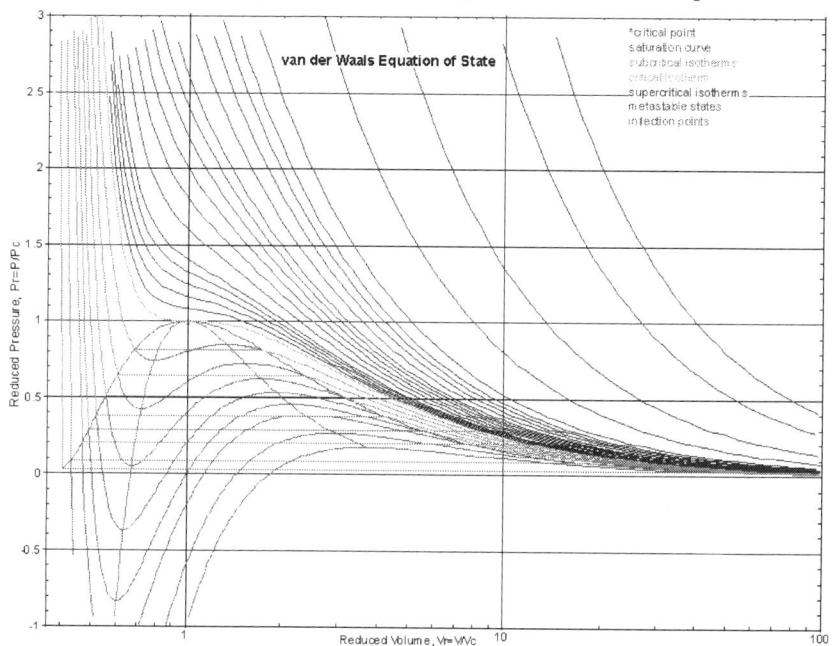

The van der Waals equation of state (i.e., an expression for pressure as a function of temperature and density or specific volume (i.e., 1/density)), $p = RT/(v-b) - a/v^2$, was the first attempt to account for observed fluid behavior. The critical point is where the red, magenta, and cyan lines meet. This is also where both the x- and y-axis equal one. The x-axis is specific volume divided by the value at the critical point and the y-axis is pressure divided by the critical pressure. The cyan curve is the locus of all points where the temperature is equal to the critical temperature. The blue curves are constant temperatures (isotherms) above the critical and the green curves are isotherms below the critical.

The red curve is the saturation line, that is, the states at which the liquid and vapor can coexist in equilibrium. The most common point for water is one atmosphere and 100°C (14.696 psia and 212°F). Notice that the green curves have a discontinuous slope where they intersect the red curve. We will spend a lot of computational time walking up and down this curve! The brown curves are metastable states, that is, they may occur, but they're not in equilibrium and won't last long. This are like water in a pot on the stove at 101°C (213°F) just before a bubble forms and rises to the surface or water in an ice tray in the freezer at -1°C (31°F) just before ice crystals form and clarity disappears.

The green curves on the left side of the critical point are very steep, which can make intersections difficult to calculate. These same curves are very shallow on the far right side, making those intersections difficult to calculate. This is a perfect illustration of nonlinear equation solving and a practical use of the bisection search and Newton-Raphson Method. All the files can be found in the online archive in the folder examples\steam. The main source code is kkhm.c and there is also a batch file to recompile it, although the executable will run on any version of the Windows® operating system.

The first test will be to calculate densities all along the red curve using these two methods. The convergence criterion is the same for both. The results are:

```
timing implicit density calculations
along the saturation line
149,636 function calls
bisection 2.359 seconds
Newton-Raphson 0.415 seconds
```

The bisection search always requires 32 iterations; whereas, the Newton-Raphson typically requires only 2 or 3. The ratio of the run times is 5.68:1, instead of 16:1, because Newton-Raphson requires the pressure and the derivative of pressure with respect to density; whereas, the bisection search does not. The 149,636 cases come from stepping along the red curve from the triple point to the critical point by 0.01°C. Some of the same summations (Equation A.4 and it's derivatives) are used to calculate the pressure and the derivative of the pressure with respect to density and we are careful to reuse these intermediate results for best efficiency. All of the functions are in kkhm.c, including the following snippet:

```
double fQ(double tau,double rho)
  {
  int i,j;
  double dt,q,r,rhoa,s,tauc;
  tauc=1000./Tc;
  dt=1.;
  rhoa=0.634;
  for(q=j=0;j<7;j++)
    {
    r=1.;
```

```
      for(s=i=0;i<8;i++)
        {
        if(j>1&&i>3)
          break;
        s+=A[7*i+j]*r;
        r*=rho-rhoa;
        }
      s+=exp(-E*rho)*(A[7*8+j]+A[7*9+j]*rho);
      q+=dt*s;
      if(j==0)
        dt=tau-tauc;
      else
        dt*=tau-2.5;
      rhoa=1.;
      }
    return(q);
    }
  double fP(double tau,double rho)
    {
    double T;
    T=1000./tau;
  return ((fQr(tau,rho)*rho+fQ(tau,rho))*rho+1.)*rho*R*T;
    }
  void rPr(double tau,double rho,double*P,double*dPdrho)
    {
    double Q,Qr,Qrr,T;
    T=1000./tau;
    Q=fQ(tau,rho);
    Qr=fQr(tau,rho);
    Qrr=fQrr(tau,rho);
    *P=((Qr*rho+Q)*rho+1.)*rho*R*T;
    *dPdrho=(((Qrr*rho+4.*Qr)*rho+2.*Q)*rho+1.)*R*T;
    }
```

It's also necessary to have some estimate of where to start the search for a root. Because we're always solving the same problem—rather than some general problem—we can use a simple curve fit to supply an estimate of liquid and vapor densities, respectively. Some software packages (including ones I've written) avoid this intensive iterative process by implementing cubic splines. Such are much faster and work reasonably well, except near the critical point, where no such method is adequate.

In fact, some of the tabulated values in the 1969 book are incorrect[19], owing (presumably) to the limited computer resources available at the time. Most the densities between 368°C and 379°C (695°F and 715°F) are off. To six significant figures the critical density is actually 0.316820 gm/cm³ (19.7784 lbm/ft³), the

---

[19] When I say "incorrect" I mean that the equations, with the constants provided, don't yield the results listed. They're close, but not close enough.

critical temperature is 647.245°K (374.095° or 705.371°F), and the critical pressure is 22.088 MPa (3203.59 psia). It is essential that the precise (and consistent!) values be used in order to satisfy the following requirement at the critical point:

$$\frac{\partial p}{\partial \rho} = \frac{\partial^2 p}{\partial \rho^2} = 0 \qquad (A.8)$$

The next thing we need is a function to return the density anywhere in the preceding figure, not just along the red curve. We will develop one with the bisection search and another using Newton-Raphson and compare the two for speed—requiring the same convergence criteria and that we *always* get a solution. Divergence or failure is *not* an option! There is a test in the code and it will stop should any of the Newton-Raphson function fail to converge.

The first step in finding the density is to determine if we're above or below the cyan curve in the preceding figure. If above, the curves are continuous and we can start searching almost anywhere. If above, the curves are discontinuous and we must limit our search to one side or the other (i.e., liquid or vapor). The curves are very steep on the liquid side (i.e., the left) so that we must not be too far off in our original estimate. The right side (vapor) isn't as much of a problem, but remember Equation A.8 because the first and second partial derivatives are both zero at the critical point. The density locating function using a bisection search is:

```
double frho1(double T,double P)  /* rho via bisection */
{
int i;
double Ps,r1,r2,rho,tau;
if(T<Tc)
   {
   Ps=Psat(T);
   if(P>Ps)
      {
      r1=1.02*frh0f(T);
      r2=P/R/T/2.5;
      }
   else
      {
      r1=P/R/T/2.5;
      r2=1.10*frh0g(T);
      }
   }
else
   {
   r1=P/R/T/2.5;
   r2=1.1;
   }
```

```
    tau=1000./T;
    for(i=0;i<32;i++)
       {
       rho=sqrt(r1*r2);
       if(fP(tau,rho)<P)
          r1=rho;
       else
          r2=rho;
       }
    return(rho);
    }
```

The functions frh0f() and frh0g() are the approximate values for saturated liquid and vapor densities, respectively. The density function using Newton-Raphson is:

```
double frho2(double T,double P)  /* rho via Newton-
   Raphson */
{
int i;
double dPdrho,eps,Pp,Ps,rho,tau;
if(T<Tc)
   {
   Ps=Psat(T);
   if(P>Ps)
      rho=frh0f(T);
   else
      rho=frh0g(T)*P/Ps;
   }
else
   rho=rhoc*(P/Pc)*(Tc/T);
if(fabs(T-Tc)>5.)
   eps=1E-6;
else
   eps=1E-4;
tau=1000./T;
for(i=0;i<99;i++)
   {
   rPr(tau,rho,&Pp,&dPdrho);
   if(fabs(Pp/P-1.)<eps)
      return(rho);
   rho+=(P-Pp)/dPdrho;
   }
fprintf(stderr,"%s(%lG,%lG) failed after %i
   iterations\n",__FUNCTION__,T,P,i);
exit(1);
}
```

Results for a series of random values of temperature and pressure are:
```
timing implicit density calculations
away from the saturation line
```

```
250,000 function calls
bisection 4.255 seconds
Newton-Raphson 1.409 seconds
```

Newton-Raphson wins out over bisection search by a factor of 3.02 this time. We will now use these density functions, along with others for enthalpy and entropy, to produce a Mollier Diagram (i.e., a graph of h vs. s), which is very important in the analysis of power systems, especially steam turbines. The figure consists of the saturation line plus a series of isobars (lines of constant pressure) and isotherms (lines of constant temperature). The necessary points are written to an Excel® spreadsheet to produce:

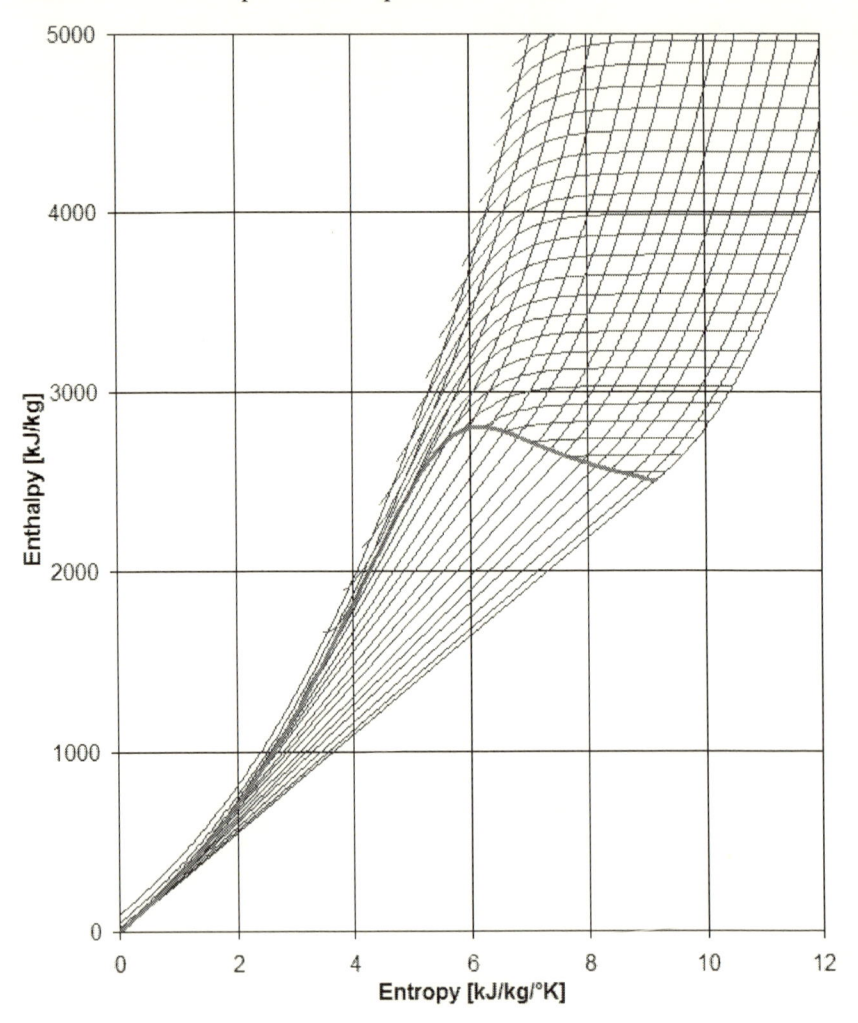

The preceding figure contains 6505 points (and far more property evaluations). The comparative performance is:

```
timing Mollier diagram calculations
bisection: 0.073 seconds
timing Mollier diagram calculations
Newton-Raphson: 0.028 seconds
```

The Newton-Raphson Method wins out over the bisection search by a factor of 2.6, which is not too impressive, considering the two combined took only a tenth of a second. The file open/close was performed outside the timer loop, but the file writes were inside. You can now see why it's tough to come up with a practical example where it's worth using anything other than a bisection search.

Appendix B: Graphical Representation

There are countless examples of mathematical problems that produce pretty pictures when rendered graphically. Unless you have some software specifically designed to facilitate this, it can be a little problematic. Excel® is not well suited for this task. All you need is a palette and a little code to convert some result into color indices. Write this out to a file and you have a pretty picture. You will find the necessary code in the online archive in folder examples\colormap. A general outline to follow is:

```
int main(int argc,char**argv,char**envp)
  {
  IterationMap(NewtonRaphson,800,800);
  SaveBitmap("Newton-Raphson.bmp");
  return(0);
  }
```

Fill an array with something, as in:

```
void IterationMap(int solver(double,double,double),int
    wide,int high)
  {
  int h,w;
  double guess,target;
  printf("creating iteration map: %ix%i\n",wide,high);
  if((iterations=calloc(wide*high,sizeof(int)))==NULL)
     Abort(__LINE__,"can't allocate memory");
  bi.biWidth=wide;
  bi.biHeight=high;
  for(h=0;h<high;h++)
     {
     guess=-3.+h*6./(high-1);
     for(w=0;w<wide;w++)
        {
        target=0.2+w*0.55/(wide-1);
        iterations[wide*h+w]=solver(target,guess,
     FLT_EPSILON);
        }
     }
  }
```

Linear variation with the number of iterations:

```
for(h=0;h<bi.biHeight;h++)
   {
   for(w=0;w<bi.biWidth;w++)
     row[w]=(BYTE)(((iterations[bi.biWidth*h+w]-
   mn)*(colors-1))/(mx-mn));
   fwrite(row,1,width,fp);
   }
```

You only need 8 bits/pixel. Don't bother with a JPEG. For a BMP the width in bytes must be a multiple of 4 (i.e., the rows must be DWORD aligned). This will take care of that:

```
bi.biBitCount=8;
width=4*((bi.biWidth*bi.biBitCount+31)/32);
```

Here's what Dekker's method from Chapter 1 looks like:

Appendix C: SIAM Paper
Applications of a Hybrid Derivative-Free Algorithm for Locating Extrema

I presented this paper at the Society of Industrial and Applied Mathematics, Southeastern Regional Seminar, held in Cullowhee, North Carolina, April 12-13, 1991. It's astounding how far computers have come in the past 27 years. Microprocessors are 100 times as fast and RAM has grown by a factor of 16. Disk space has increased ten-fold (2^{10}=1024). The test cases described herein are included in the online archive. They are still of interest, although there is less motivation to develop efficient methods with faster processing.

Abstract

Applications of a hybrid derivative-free algorithm for locating extrema of nonlinear functions of several variables based on Broyden's method is presented in which the problems of starting values and extraneous entrapment are addressed. The principal intended application of the algorithm is to find solutions to simultaneous nonlinear equations. The main objective of the algorithm is to minimize the number of function evaluations for problems where the equations are computationally intensive or partial derivatives cannot be determined analytically. Four examples drawn from diverse fields are given for illustration. Comparisons are made to the Newton-Raphson, conjugate-gradient, and steepest-descent methods.

Nomenclature

A=rectangular matrix having M columns and N rows
B=column matrix having M elements
F=a function of several variables
M=the number of residuals ($M \geq N$)
N=the number of unknowns
R=residual column matrix having M elements
X=unknown column matrix having N elements
superscript
T=matrix transpose
subscripts
N=new or current value
O=old or previous value

Introduction

Many practical problems can be cast into the form of a search for extrema of a function of several variables. A common function is the sum of squared residuals, in which case the extrema of interest are the roots of simultaneous equations.

Methods abound which require knowledge of the partial derivatives. Many of these derivative-based methods can be adapted by using finite differences to solve problems where the partial derivatives cannot be analytically determined. Such implementations are impractical when the function is computationally intensive.

Derivative-based methods such as the Newton-Raphson discard at each step all information previously learned about the behavior of the function except the current location. Even the Conjugate-Gradient method when applied to nonlinear problems may only preserve one previous direction of search. When the function evaluation is computationally intensive it is essential that as much information as possible about the behavior of the function learned from previous evaluations be preserved and utilized.

Broyden's method is very attractive when considered from this perspective. It does not require knowledge of the partial derivatives, nor does it attempt to compute them directly. Furthermore, Broyden's method preserves all of the information learned about the behavior of the function for the last $N+1$ steps where N is the number of unknowns.

Four enhancements to Broyden's method were made to arrive at the present algorithm: a method for selecting starting values, step length control, hybrid search algorithm, and a method for escaping from extraneous entrapment.

The Basic Method

Given a set of N unknowns represented by the column matrix X and a corresponding set of M residuals represented by the column matrix R, the least-squares function would be $F=R^T R$. The extrema of F occur at the locations where $MF/MX=0$. If the residuals, R, were linear functions of the unknowns, X, then the function, F, would be quadratic and its contours would plot as ellipsoids. This linear case could be described by Equations C.1 and C.2:

$$R = A^T X + B \qquad (C.1)$$

$$F = R^T R = X^T A A^T X + 2 B^T A^T X + B^T B \qquad (C.2)$$

where A is a rectangular matrix having M columns and N rows and B is a column matrix having M elements.

Broyden (1969) reasoned that A and B should be selected such that exact agreement would be preserved for the previous $N+1$ steps. Assuming that no two of the previous $N+1$ Xs are the same, there should be a unique solution to the resulting $M(N+1)$ equations for the elements of A and B. Ignoring for the moment how this sequence of Xs might be obtained, the matrices A and B can be sequentially updated using the following algorithm:

$$A_N = \frac{A_O + [(R_N - R_O) - A_O(X_N - X_O)](X_N - X_O)^T}{(X_N - X_O)^T(X_N - X_O)} \quad \text{(C.3)}$$

$$B_N = R_N - A_N^T X_N \quad \text{(C.4)}$$

where the subscripts N and O refer to *new* and *old* respectively—or the current step and the previous one. Equations C.3 and C.4 can be verified by substitution into Equation C.1 with the *new* and *old* subscripts added. If A and B are initialized to zero and $N+1$ unique starting values of X are selected, then after $N+1$ function evaluations and updates, matrices A and B will be uniquely defined and the search for a solution could proceed.

The interesting property of Equation C.3, which led Broyden to this selection is that, the change in A is only in the direction of the last step in X. That is, the only information about the behavior of the function, which is added to A at each step, is its variation along the current search direction. All of the information about the function in the $N-1$ directions orthogonal to the current search direction remains intact; thus, it is a *rank-one* update method.

Broyden used this algorithm for obtaining and updating matrices A and B, along with Newton's method to search for the extrema. Thus in its original form, Broyden's is a quasi-Newton method (Morè and Sorensen (1984) discuss Newton and quasi-Newton methods in some detail.). The following calculus can be applied to the matrices in order to illustrate this procedure.

$$\frac{\partial R}{\partial X} = A^T \quad \text{(C.5)}$$

$$R_O = A_O^T(X_O - X_N) \quad \text{(C.6)}$$

Matrix A contains the partial derivatives of the residuals, R, with respect to the unknowns, X. Thus matrix A is the *Jacobian* of R with respect to X.

As indicated by Equation C.6, the gradient of the function lies along the direction AR; therefore, most any gradient search method could be implemented and updated using Broyden's method for determining the Jacobian. Ortega and Rheinboldt (1970) discuss on an analytical level a number of methods which could be applied at this point. Actual selection of a practical method, which will produce satisfactory stable results for a wide range of problems, is quite another matter.

The Modified Method

Nonlinear simultaneous equations may have no solution, one solution, or many solutions. The most helpful physical analogy is that of a relief map of the Earth's surface where the unknowns are latitude and longitude and the function is the elevation with respect to mean sea level. No conceivable practical method could hope to locate Mt. Everest or the Marianas Trench regardless of the starting values. While it is reasonable to search for local extrema, it is fortuitous to locate the global extremum—assuming one does exist. Given this analogy it is understandable that no practical algorithm can be expected to locate even a local extremum in every case. Fletcher (1987) discusses these and other problems related to locating extrema in more detail.

Selection of Starting Values

This geographical analogy illustrates the necessity of limiting the region to be searched for extrema. In the present algorithm, a minimum and maximum value for each element in X must be supplied. This not only provides an extent to the range of X, but it also serves as an indication of the scale. Any change in X, which is on the order of the machine precision when compared to the range of X, is considered negligible. One logical choice for the $N+1$ starting values of X would be the center plus the N evenly distributed surrounding values inside the hypercube defined by the specified range of X.

If the function at the central point is greater than at the surrounding points, then the first iteration would direct the search outside of the range of X. If this occurs the range is bisected such that the new center point is mid way between the previous center and the surrounding point corresponding to the least value of the function. If this bisection is unsuccessful after sufficient attempts so as to diminish the subrange of any element of X to the previously determined negligible level, the search is abandoned.

Step Length Control

The unmodified method often results in unstable iterations. Not only is it necessary to confine X to the specified range, it is also necessary to damp the iteration or, as in this case, apply a step length control algorithm. Ortega and Rheinboldt discuss several step length algorithms. The parabola method defined by the current location, one *close* point, and the next step prescribed by the unmodified method has proven to be as successful as any tested. Using the unmodified point as an outer limit on the step length arises from the observation that the unmodified method has a strong tendency to overshoot.

Hybrid Search Algorithm

Because Broyden's is basically a quasi-Newton method, the search proceeds in much the same direction as with *Newton-Raphson* (*NR*). In cases where the *NR* method would fail to locate an extremum, most likely Broyden would also. Broyden's method can also be viewed as a means by which to obtain the Jacobian (matrix A). If the Newton iteration is not successful, the Jacobian can be used to implement other methods. The method of *steepest descent* (*SD*) is more robust, but converges less rapidly than *NR*. When the *NR* iteration fails to result in a reduction of the residual, the direction of steepest descent is searched.

In the present algorithm, the *Conjugate-Gradient* (*CG*) method with the restart procedure recommended by Powell (1977) is also used to supplement the *NR/SD* iteration. The only information added to the Jacobian by Broyden's update is along the search direction. Information about the character of nonlinear functions in directions orthogonal to the search direction can be essential to locating extrema. The *CG* method provides a systematic procedure for searching other directions. In the present algorithm, the *NR*, *SD*, and *CG* methods are used alternately as each ceases to provide continual reduction of the residual.

Escape from Extraneous Entrapment

If N directions have been searched without further improvement, then either a local extremum has been found or extraneous entrapment has occurred. Whether the current location is a local extremum or a nuisance of finite-precision arithmetic can be partially determined by examining the history of matrix A. For nonlinear problems the character of A can change substantially as the search proceeds.

The unmodified Broyden update to A replaces the information along the direction of the current step—thus discarding the previous information along this direction. If an *old* copy of A is retained along with the *new* copy and the search direction indicated by the old A is away from that indicated by the new A (viz. the dot product of the column matrices is less than or equal to zero), then the iteration may have skipped over an inflection point. In this case a search is conducted along the direction connecting these two provisional new values of X.

Extraneous entrapment can sometimes be corrected by arbitrarily perturbing the solution away from the current location to see if it will return to the same point. After this perturbation has been attempted without success in N directions the procedure is abandoned.

Extension to Least Squares

In the case where $M>N$, Equation C.7 must be pre-multiplied by A. The simultaneous nonlinear equations are then solved in the least-squares sense. For most problems the stability of the method also improves when this multiplication is performed even in the case of $M=N$. Therefore, in the present algorithm it is done regardless of the values of N and M.

Comparison to Other Methods

The present derivative-free *enhanced* Broyden (*EB*) method was compared to the Newton-Raphson (*NR*) and Conjugate-Gradient (*CG*) methods. The results are listed in Table 1. All three methods have step-size control and for the test cases were required to obtain essentially the same solution. All three methods were given the same starting values (initial guess) so that there was no advantage of one over the other in these respects.

Table 1 lists the number of variables (independent unknowns and dependent residuals), the number of function evaluations, and relative performance. The relative performance is the number of CPU-seconds required for the *NR* divided by the number required for the particular method (thus, *NR* will always have a relative performance of 1.0).

Test Case 1

The first test case is a nonlinear constrained curve-fitting problem. The best fitting single branch of a hyperbola was sought which would not only agree with the data (in this case experimental film boiling droplet area as a function of time), but would also have asymptotic characteristics conforming to the observed phenomena. The resulting curve fit must have one and only one root. The root must lie outside the range of the data and the derivative must be infinite at that point. The problem is nonlinear because of the constraints and the form (a rational polynomial). The partial derivatives of the residual cannot be determined analytically as these result in yet another set of simultaneous nonlinear equations. This test case was selected as being typical from among a set of 125.

Test Case 2

The second test case is similar to a nonlinear unconstrained curve-fitting problem. The values of hydraulic conductivity and storativity (groundwater analogs of electrical conductance and capacitance) were sought which would best characterize a measured field response. A field test was conducted by pumping water from a well and measuring the change in the water table in a nearby well. An analytical expression for the ideal response of an aquifer contains these two unknown parameters, which must be selected so as to best agree with the measured response. This problem is nonlinear, however the partial derivatives of the residual can be computed analytically (see note * in the table). This test case was selected as being typical from among a set of 33.

Test Case 3

The third test case is the determination of four *calibration factors* (mass transfer and pressure drop coefficients characterizing a particular type of plastic media), which are needed to run a large finite-integral code (numerical model of a cooling tower). Forty-nine sets of field data were collected for this plastic media. What were sought are the calibration factors which when input to the model would best reproduce the measured results. The finite-integral code itself was repeatedly run to provide the residuals. Needless to say, this was a very computationally intensive process—one in which minimizing the number of function evaluations was crucial. This test case, which is actually a type of inverse mass transfer problem, was selected as being typical from among a set of 6.

Test Case 4

The fourth test case is the determination of 4 phase lags and 4 corresponding weights, which would best characterize the transient response of a dammed reservoir. A linear model was sought for the cross-sectionally averaged transient flow at a specific location (adjacent to a large power plant) within a reservoir bounded by two dams, which are used for peaking (i.e., they discharge water only during times of peak electrical demand). This linear model was to become part of a larger linear systems optimization code used for long-range planning and resource management. An existing dynamic fluid flow model was used along with historical dam operations to produce a target data set. This test case was selected as being typical from among a set of 4.

Discussion

The focus of these four test cases is not on many variables, but on non-analytically differentiable residuals and lengthy function evaluations. In each case there is some physical phenomenon, which provides the basis for the residuals. Because these test cases are based on physical phenomena, the approximate bounds on the solution are also known. In each case lengthy graphical or cumbersome numerical techniques exist for finding extrema. The advantages to using the present algorithm in these cases are convenience and speed.

Table 1. Comparison of Methods for Locating Extrema

test case	variables		function evaluations			performance index		
	N	M	N-R	C-G	E-B	N-R	C-G	E-B
1	4	50	417	462	76	1	0.8	1.9
2	2	47	35*	51*	61	1	0.8	3.1
3	4	98	137	298	14	1	0.5	10.4
4	8	8760	103	344	18	1	0.4	2.4

*Note: Functions calls reporting analytical partial derivatives are more time consuming than those, which do not.

In the first test case (fitting a hyperbola with constraints and later taking its derivative) has been done for years using hand-drawn curves and a drafting protractor. The second test case (determining hydraulic conductance and capacitance) has also been done for years by graphical means and more recently by asymptotic extension to separate the coupled influence of the unknowns. The third test case (determining calibration factors for mass transfer and pressure drop) has typically been done by assuming half of the unknowns to be the same as a similar media and computing the others by *trial-and-error*.

For these test cases the average performance of the *EB* method is about 4 times the *NR* and *CG* methods. As mentioned previously, these are not isolated examples, but *working problems* from a variety of fields, which were the impetus for developing the method. The *EB* method utilizes the best features of the *NR*, *CG*, and *SD* methods along with avoiding direct calculation of the Jacobian. The relative advantage of the *EB* method was most dramatic for Test Case 3 where the difference in runtime was a matter of days (on a 33MHz-80386/7 machine).

A two-variable problem is best suited to illustrate the searching procedure graphically. Figures 1 through 3 show the contours of the function in Test Case 2 and the first few steps in the search path for the *NR*, *CG*, and *EB* methods respectively. The Z-axis or the contours is percent total residual (in 20% intervals). The dark (dense dot) region is close to the extremum and the light (sparse dot) region is far from the extremum. This graphical format was selected in order to give a *bulls-eye* appearance.

In this case the *CG* essentially follows the gradient inward to the center of the *bulls-eye* (see the dark path line in Figure 2). The *CG* path is almost perpendicular to the contours as it crosses each one. The *SD* path if it were shown would differ little from the *CG*. The *NR* and *EB* paths differ markedly from the *CG* (compare the dark path lines in Figures 1 and 3 to Figure 2). The *NR* and *EB* methods reach the vicinity of the extremum (i.e., penetrate the darkest inner contour) in significantly fewer steps than does the *CG* method.

The hybrid implementation of the present method can be seen by comparing the second step in the *NR* and *EB* paths. The line connecting the second and third points on the *NR* line (Figure 1) is almost parallel to the contour next to it (i.e., this step is almost perpendicular to the gradient). The line connecting the second and third points on the *EB* path (Figure 3) is almost perpendicular to the contour (i.e., almost in line with the gradient at the point where it crosses the inner contour). This illustrates how the *EB* method checks the search direction corresponding to all three methods (*NR*, *CG*, and *SD*) to see which is more advantageous at a particular location.

Conclusions

Broyden's derivative-free method for solving nonlinear simultaneous equations has been presented along with four enhancements. These enhancements include: a method for selecting starting values, step length control, hybrid search algorithm, and a method for escaping from extraneous entrapment. A significant performance improvement over the Newton-Raphson and Conjugate-Gradient methods is shown for four test cases taken from varied fields. Part of this performance improvement is a consequence of the derivative-free method. The hybrid search algorithm used in this enhanced Broyden method further improves the performance by utilizing the strengths of three other methods (the Newton-Raphson, Conjugate-Gradient, and Steepest-Descent).

References

Broyden, C., "A New Method of Solving Nonlinear Simultaneous Equations," *Computational Journal*, Vol. 12, pp. 94-99, 1969.

Fletcher, R., *Practical Methods of Optimization*, John Wiley and Sons, New York, NY, 1987.

Morè, J. J. and D. C. Sorensen, "Newton's Method," *Studies in Numerical Analysis*, G. H. Golub, ed., The Mathematical Association of America, pp. 29-82, 1984.

Ortega, J. M. and W. C. Rheinboldt, *Iterative Solution of Nonlinear Equations in Several Variables*, Academic Press, New York, 1970.

Powell, M. J. D., "Restart Procedures for the Conjugate Gradient Method," *Mathematical Programming*, Vol. 12, pp. 241-254, 1977.

also by D. James Benton

3D Rendering in Windows: How to display three-dimensional objects in Windows with and without OpenGL, ISBN-9781520339610, Amazon, 2016.

Curve-Fitting: The Science and Art of Approximation, ISBN-9781520339542, Amazon, 2016.

Differential Equations: Numerical Methods for Solving, ISBN-9781983004162, Amazon, 2018.

Evaporative Cooling: The Science of Beating the Heat, ISBN-9781520913346, Amazon, 2017.

Heat Exchangers: Performance Prediction & Evaluation, ISBN-9781973589327, Amazon, 2017.

Jamie2 2nd Ed.: Innocence is easily lost and cannot be restored, ISBN-9781520339375, Amazon, 2016-18.

Little Star 2nd Ed.: God doesn't do things the way we expect Him to. He's better than that! ISBN-9781520338903, Amazon, 2015-17.

Living Math: Seeing mathematics in every day life (and appreciating it more too), ISBN-9781520336992, Amazon, 2016.

Lost Cause: If only history could be changed…, ISBN-9781521173770. Amazon 2017.

Mill Town Destiny: The Hand of Providence brought them together to rescue the mill, the town, and each other, ISBN-9781520864679, Amazon, 2017.

Monte Carlo Simulation: The Art of Random Process Characterization, ISBN-9781980577874, Amazon, 2018.

Numerical Calculus: Differentiation and Integration, ISBN-9781980680901, Amazon, 2018.

ROFL: Rolling on the Floor Laughing, ISBN-9781973300007, Amazon, 2017.

A Synergy of Short Stories: The whole may be greater than the sum of the parts, ISBN-9781520340319, Amazon, 2016.

Thermodynamics - Theory & Practice: The science of energy and power, ISBN-9781520339795, Amazon, 2016.

Version-Independent Programming: Code Development Guidelines for the Windows® Operating System, ISBN-9781520339146, Amazon, 2016.

Made in the USA
Middletown, DE
14 April 2021